Groundwater Quality and the Relation Between pH Values and Occurrence of Trace Elements and Radionuclides in Water Samples Collected from Private Wells in Part of the Kickapoo Tribe of Oklahoma Jurisdictional Area, Central Oklahoma, 2011

By Carol J. Becker

Prepared in cooperation with the Kickapoo Tribe of Oklahoma

Scientific Investigations Report 2012–5253
Revised March 2013

U.S. Department of the Interior
U.S. Geological Survey

U.S. Department of the Interior
KEN SALAZAR, Secretary

U.S. Geological Survey
Suzette M. Kimball, Acting Director

U.S. Geological Survey, Reston, Virginia: 2013

This and other USGS information products are available at http://store.usgs.gov/
U.S. Geological Survey
Box 25286, Denver Federal Center
Denver, CO 80225

Suggested citation:
Becker, C.J., 2013, Groundwater quality and the relation between pH values and occurrence of trace elements and radionuclides in water samples collected from private wells in part of the Kickapoo Tribe of Oklahoma Jurisdictional Area, central Oklahoma, 2011: U.S. Geological Survey Scientific Investigations Report 2012–5253, 47 p., 5 apps. (Revised March 2013.)

Acknowledgments

The author thanks Darren Shields, Kara Davis, and Richard Ketakeah of the Kickapoo Tribe of Oklahoma for their assistance with this project. Gratitude also is expressed to the many well owners who graciously allowed the author to access their wells for collection of water samples.

The author would like to thank U.S. Geological Survey (USGS) personnel Zoltan Szabo and Lisa Senior for reviewing the report draft and providing constructive comments. Mr. Szabo's assistance with interpreting the radionuclide quality-control sample analyses was essential. The author would also like to thank USGS personnel Ann Mullin at the USGS National Water Quality Laboratory for her assistance with analysis and interpretation of the radionuclide samples.

Contents

Figures

Tables

Conversion Factors

Inch/Pound to SI

Multiply	By	To obtain
Length		
inch (in.)	2.54	centimeter (cm)
inch (in.)	25.4	millimeter (mm)
foot (ft)	0.3048	meter (m)
mile (mi)	1.609	kilometer (km)
Area		
square mile (mi^2)	259.0	hectare (ha)
square mile (mi^2)	2.590	square kilometer (km^2)
Volume		
gallon (gal)	3.785	liter (L)
gallon (gal)	0.003785	cubic meter (m^3)
gallon (gal)	3.785	cubic decimeter (dm^3)
ounce, fluid (fl. oz)	0.02957	liter (L)
ounce, fluid (fl. oz)	29.574	milliliter (mL)

Temperature in degrees Celsius (°C) may be converted to degrees Fahrenheit (°F) as follows:

°F=(1.8×°C)+32

Temperature in degrees Fahrenheit (°F) may be converted to degrees Celsius (°C) as follows:

°C=(°F-32)/1.8

Vertical coordinate information is referenced to the North American Vertical Datum of 1988 (NAVD 88).

Horizontal coordinate information is referenced to the North American Datum of 1983 (NAD 83).

Altitude, as used in this report, refers to distance above the vertical datum.

Specific conductance is given in microsiemens per centimeter at 25 degrees Celsius (μS/cm at 25 °C).

Concentrations of chemical constituents in water are given either in milligrams per liter (mg/L) or in micrograms per liter (μg/L).

Concentrations of radionuclides in water are given in picocuries per liter (pCi/L).

Laboratory reporting level (LRL): The LRL is generally equal to twice the yearly determined long-term method detection level (LT–MDL). The LRL controls false negative error. The probability of falsely reporting a nondetection for a sample that contained an analyte at a concentration equal to or greater than the LRL is predicted to be less than or equal to 1 percent. The value of the LRL will be reported with a "less than" remark code for samples in which the analyte was not detected. The National Water Quality Laboratory collects quality-control data from selected analytical methods on a continuing basis to determine LT–MDLs and to establish LRLs. These values are reevaluated annually based on the most current quality-control data, and therefore, may change (Childress and others, 1999).

Sample-specific critical level (ssLc): The critical level is the lowest measured concentration that is statistically different from the instrument background or analytical blank, and it serves as the detection threshold for deciding whether the radionuclide is present in a sample (McCurdy and others, 2008).

Groundwater Quality and the Relation Between pH Values and Occurrence of Trace Elements and Radionuclides in Water Samples Collected from Private Wells in Part of the Kickapoo Tribe of Oklahoma Jurisdictional Area, Central Oklahoma, 2011

By Carol J. Becker

Abstract

From 1999 to 2007, the Indian Health Service reported that gross alpha-particle activities and concentrations of uranium exceeded the Maximum Contaminant Levels for public drinking-water supplies in water samples from six private wells and two test wells in a rural residential neighborhood in the Kickapoo Tribe of Oklahoma Jurisdictional Area, in central Oklahoma. Residents in this rural area use groundwater from Quaternary-aged terrace deposits and the Permian-aged Garber–Wellington aquifer for domestic purposes. Uranium and other trace elements, specifically arsenic, chromium, and selenium, occur naturally in rocks composing the Garber–Wellington aquifer and in low concentrations in groundwater throughout its extent. Previous studies have shown that pH values above 8.0 from cation-exchange processes in the aquifer cause selected metals such as arsenic, chromium, selenium, and uranium to desorb (if present) from mineral surfaces and become mobile in water. On the basis of this information, the U.S. Geological Survey, in cooperation with the Kickapoo Tribe of Oklahoma, conducted a study in 2011 to describe the occurrence of selected trace elements and radionuclides in groundwater and to determine if pH could be used as a surrogate for laboratory analysis to quickly and inexpensively identify wells that might contain high concentrations of uranium and other trace elements.

The pH and specific conductance of groundwater from 59 private wells were measured in the field in an area of about 18 square miles in Lincoln and Pottawatomie Counties. Twenty of the 59 wells also were sampled for dissolved concentrations of major ions, trace elements, gross alpha-particle and gross beta-particle activities, uranium, radium-226, radium-228, and radon-222 gas.

Arsenic concentrations exceeded the Maximum Contaminant Level of 10 micrograms per liter in one sample having a concentration of 24.7 micrograms per liter. Selenium concentrations exceeded the Maximum Contaminant Level of 50 micrograms per liter in one sample having a concentration of 147 micrograms per liter. Both samples had alkaline pH values, 8.0 and 8.4, respectively. Uranium concentrations ranged from 0.02 to 383 micrograms per liter with 5 of 20 samples exceeding the Maximum Contaminant Level of 30 micrograms per liter; the five wells with uranium concentrations exceeding 30 micrograms per liter had pH values ranging from 8.0 to 8.5. Concentrations of uranium and radon-222 and gross alpha-particle activity showed a positive relation to pH, with the highest concentrations and activity in samples having pH values of 8.0 or above. The groundwater samples contained dissolved oxygen and high concentrations of bicarbonate; these characteristics are also factors in increasing uranium solubility.

Concentrations of radium-226 and radium-228 (combined) ranged from 0.03 to 1.7 picocuries per liter, with a median concentration of 0.45 picocuries per liter for all samples. Radon-222 concentrations ranged from 95 to 3,600 picocuries per liter with a median concentration of 261 picocuries per liter. Eight samples having pH values ranging from 8.0 to 8.7 exceeded the proposed Maximum Contaminant Level of 300 picocuries per liter for radon-222. Eight samples exceeded the 15 picocuries per liter Maximum Contaminant Level for gross alpha-particle activity at 72 hours (after sample collection) and at 30 days (after the initial count); those samples had pH values ranging from 8.0 to 8.5. Gross beta-particle activity increased in 15 of 21 samples during the interval from 72 hours to 30 days. The increase in gross beta-particle activity over time probably was caused by the ingrowth and decay of uranium daughter products that emit beta particles.

Water-quality data collected for this study indicate that pH values above 8.0 are associated with potentially high concentrations of uranium and radon-222 and high gross alpha-particle activity in the study area. High pH values also are associated with potentially high concentrations of arsenic, chromium, and selenium in groundwater when these elements occur in the aquifer matrix along groundwater-flow paths.

Introduction

From 1999 to 2007, the Indian Health Service reported that gross alpha-particle activities and concentrations of uranium exceeded the Maximum Contaminant Levels (MCL) for public drinking-water supplies in water samples from six private wells and two test wells in a rural residential neighborhood in the Kickapoo Tribe of Oklahoma Jurisdictional Area, in central Oklahoma (D. Shields, Director, Department of Environmental Programs, Kickapoo Tribe of Oklahoma, written commun., 2010). Gross alpha-particle activities in water samples from the private wells were reported to range from 20 to 789 picocuries per liter (pCi/L), exceeding the MCL of 15 pCi/L and two of the six wells contained uranium concentrations of 46 and 47 micrograms per liter (µg/L), exceeding the MCL of 30 µg/L (U.S. Environmental Protection Agency, 2009a). Water samples from two test wells drilled in the neighborhood to obtain additional water-quality data contained gross alpha-particle activities of 41 and 58 pCi/L and uranium concentrations of 38 to 51 µg/L. A water sample collected from one of the six private wells by the U.S. Geological Survey (USGS) in 2008 contained a uranium concentration of 1,500 µg/L. The pH values of that sample and water from the two test wells were alkaline at 8.2, 8.6, and 8.9 standard units, respectively. These private wells have been subsequently destroyed but because the quality of water from private wells is not routinely measured, other well owners in the area may not be aware of water-quality problems that can increase risks to their health.

Residents in this rural area use groundwater from Quaternary-aged terrace deposits and the Permian-aged Garber–Wellington aquifer for domestic purposes. The Garber–Wellington aquifer is part of a complex of sandstone, siltstone, and mudstone rocks, referred to as the "Central Oklahoma aquifer," that underlies about 2,890 square miles of central Oklahoma (fig. 1) and is the primary water supply for municipal, domestic, industrial, and agricultural needs (Tortorelli, 2009). Uranium, arsenic, chromium, and selenium occur naturally in rocks in the Garber Sandstone and the Wellington Formation (that compose the Garber–Wellington aquifer) and in low concentrations in groundwater throughout the extent of those units (Christenson and Havens, 1998; Schlottmann and others, 1998). In some areas, concentrations of these trace elements exceed the MCLs and have created challenges for public water-supply systems that must provide water below regulated concentration levels to the public (Schlottmann and others, 1998). In central Oklahoma, arsenic is the trace element of highest concern because its occurrence in concentrations above drinking-water standards affects more public-water supply systems than does any other inorganic constituent in groundwater (Schlottmann and others, 1998; Robertson, 1989; Parkhurst and others, 1996; Smith and others, 2009). The extent and occurrence of elevated concentrations of uranium and other trace elements in private well water are relatively unknown in the Kickapoo Tribe of

Oklahoma Jurisdictional Area because of the relative lack of water-quality data from private wells compared to water used for public supply.

On the basis of these previously collected data, the USGS, in cooperation with the Kickapoo Tribe of Oklahoma, sampled groundwater in an area of about 18 square miles in Lincoln and Pottawatomie counties in the Kickapoo Tribe of Oklahoma Jurisdictional Area, to describe the occurrence of selected trace elements and radionuclides in groundwater and to determine if pH could be used as a surrogate for more costly analyses to quickly and inexpensively identify groundwater that might contain high concentrations of selected trace elements and radionuclides. Data described in this report may be useful for making decisions about future well-drilling activities and for prioritizing areas for water-quality testing, water treatment, water use, and other planning purposes.

Purpose and Scope

This report describes the results of a study conducted in 2011 to describe groundwater quality and the relation between pH values and the occurrence of selected trace elements and radionuclides in samples collected from private wells in an area of about 18 square miles in Lincoln and Pottawatomie Counties in the Kickapoo Tribe of Oklahoma Jurisdictional Area in central Oklahoma (fig. 1). Field measurements of pH and specific conductance of water from 59 private wells are summarized, and the results of chemical analyses of water from 20 of the 59 wells selected for sampling are presented and summarized. Historical pH values and concentrations of arsenic and uranium in the Kickapoo Tribe of Oklahoma Jurisdictional Area are shown on maps and graphs for comparison to collected samples. These data were retrieved from the USGS National Water Information System database on December 1, 2011 (U.S. Geological Survey, 2012).

Health Information on Radionuclides and Radioactivity

Uranium-238 and thorium-232 are the most common naturally occurring radioactive elements and are found in varying amounts in soils, rocks, and water (Zapecza and Szabo, 1988; Focazio and others, 2001). Both of these radionuclides are unstable and undergo a decay process forming progeny ("daughter") radionuclides to stabilize the atomic configuration; those daughter radionuclides undergo further radioactive decay (fig. 2). Some elements have multiple isotopes with each isotope having the same number of protons and electrons but differing numbers of neutrons. The different number of neutrons gives each isotope of an element a different mass number, which affects the nucleus stability and chemical bond strength. For example, there are 17 isotopes of uranium, of which 3 are naturally occurring (uranium-238, uranium-235, and uranium-234). All uranium isotopes are radioactive, and most decay in fractions of a second.

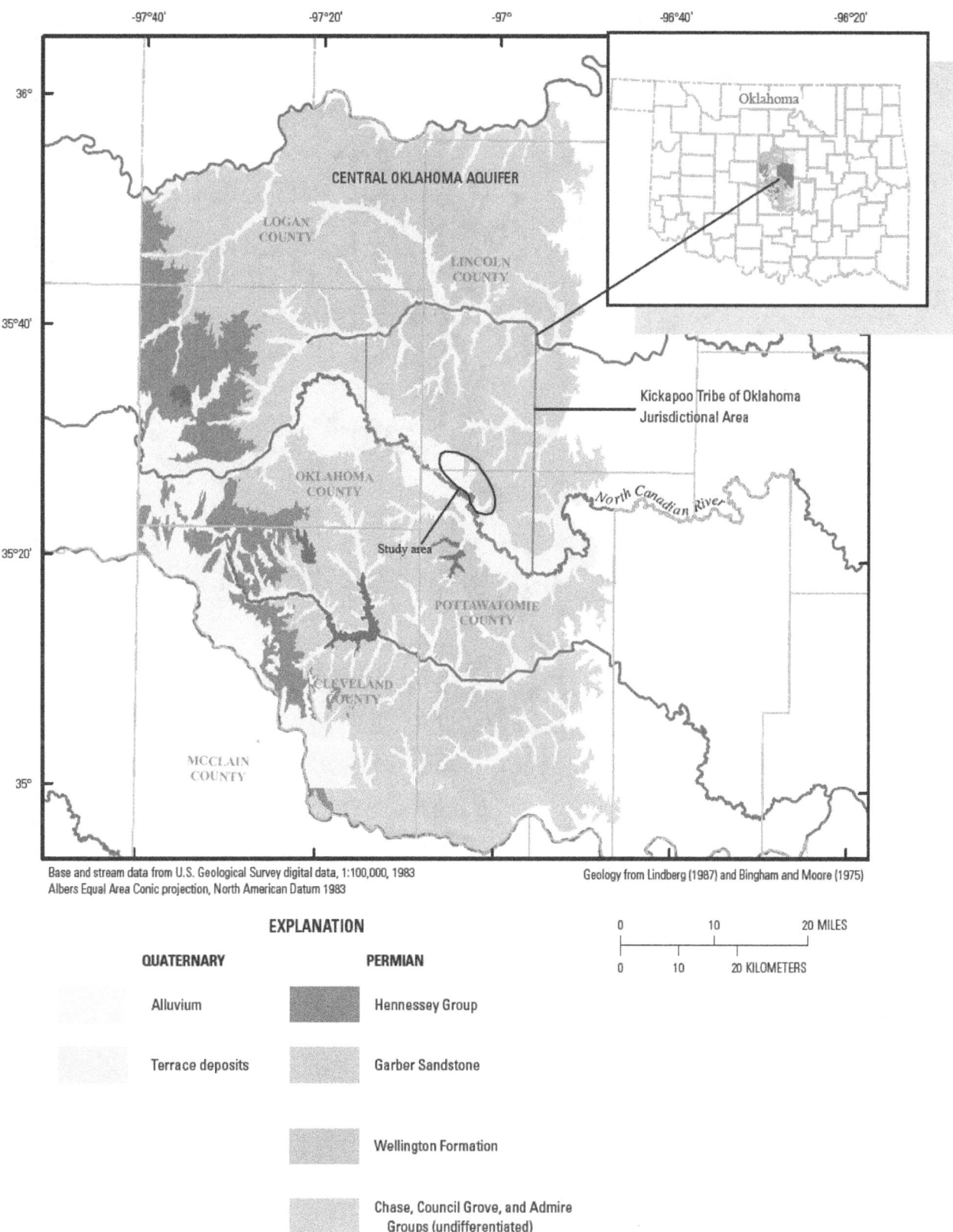

Figure 1.   Surficial geology and location of the Central Oklahoma aquifer and study area in the Kickapoo Tribe of Oklahoma Jurisdictional Area, central Oklahoma.

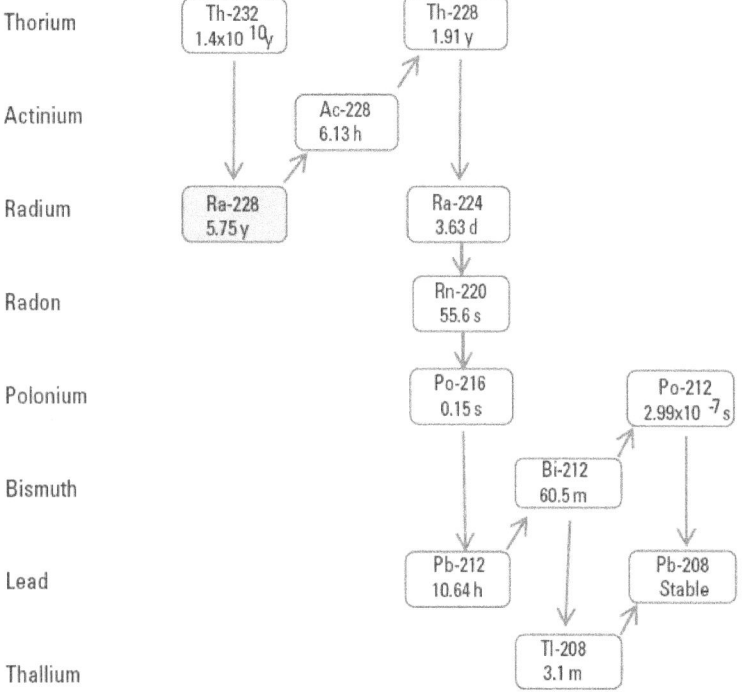

Figure 2.   Uranium-238 and thorium-232 radioactive decay series.

The uranium-238 isotope is the most stable (long-lived) uranium isotope, having a half-life of about 4.5 billion years; as a result it composes 99.4 percent of the uranium in the Earth (a half-life is the time required for a radioactive element to decay to half of the initial amount). This isotope is sufficiently abundant to be detected by relatively common chemical analyses.

The radioactive decay process releases ionizing radiation, which is composed of alpha particles, beta particles, or gamma rays. Alpha particles do not penetrate the skin and are hazardous to internal organs only when inhaled or ingested (U.S. Environmental Protection Agency, 1999). Beta particles can penetrate the skin, and strong sources can redden or even injure tissue; however, the greatest damage occurs through inhalation or ingestion (U.S. Environmental Protection Agency, 2012a). The greatest risk to human health in general comes from a lifetime of exposure to radionuclides (U.S. Environmental Protection Agency, 1999); health hazards from short-term exposure generally are negligible. Drinking water MCLs for exposure to radioactivity are set at a level of about a 1 in 10,000 risk of developing a fatal cancer over a lifetime of 70 years when consuming 2 liters of water per day (U.S. Environmental Protection Agency, 2000).

Uranium as the isotope uranium-238 decays through intermediate decay products to radium-226 (half-life of 1,601 years), which decays to radon-222 gas (half-life of 3.82 days) (fig. 2). Uranium isotopes, radium-226, and radon-222 are primarily alpha-particle emitters but have short-lived intermediate progenies created during the decay process that emit beta particles (Zapecza and Szabo, 1988). The main health effects from long-term ingestion of uranium are kidney damage and an increase in the risk of cancer (U.S. Environmental Protection Agency, 2011). Uranium has an MCL in public drinking-water supplies of 30 μg/L (U.S. Environmental Protection Agency, 2000).

Radium-228, a daughter product of the most abundant isotope of thorium-232, has a half-life of 5.75 years and emits beta particles during the decay process. Health effects from long-term exposure to radium isotopes by ingestion are an increase in the risk of cancer and diseases that affect the formation of blood such as leukemia and aplastic anemia (U.S. Environmental Protection Agency, 2010a). Radium-226 and radium-228 in public drinking-water supplies have a combined MCL of 5 pCi/L.

Radon-222 is the most abundant and longest lived radon isotope. Radon occurs as a gas that can be dissolved in water and ingested, but inhalation is the primary method of exposing the body to radiation (U.S. Environmental Protection Agency, 2010b). Radon gas dissolved in water can be released indoors when one showers or washes, but the main source of radon emission to indoor air is the decay of uranium in soils and rocks underlying homes and buildings (U.S. Environmental Protection Agency, 2010b). Radon is the second leading cause of lung cancer in the United States and is considered to be a serious potential health problem (Commission on Life Sciences, 1999). As of 2012, there was no MCL for radon in public drinking-water supplies. The MCL of 4,000 pCi/L for radon in public water-supply systems has been proposed by the U.S. Environmental Protection Agency (EPA) if statewide Multimedia Mitigation programs are established to address radon in indoor air. A second option is a MCL of 300 pCi/L for systems that choose not to develop Multimedia Mitigation programs (U.S. Environmental Protection Agency, 2012b).

Study Area Description

The Kickapoo Tribe of Oklahoma Jurisdictional Area encompasses about 317 square miles in parts of Lincoln, Oklahoma, and Pottawatomie Counties (fig. 1). The study area was located in the south-central portion of the Kickapoo Tribe of Oklahoma Jurisdictional Area over an area of about 18 square miles. The area is mostly rural with households dependent on wells for water supply and septic systems for sewage. Land cover is predominately grassland (70 percent) with about 10 percent used for hay and pasture and 10 percent for cultivated crops, which are located on the alluvium (Fry and others, 2011).

Private wells in the study area range in depth from less than 100 feet (ft) to generally not more than 200 ft below land surface and obtain water from Quaternary-aged terrace deposits and the underlying Permian-aged Garber–Wellington aquifer (Oklahoma Water Resources Board, 2011). Wells completed in the adjoining alluvium of the North Canadian River (fig. 1) generally are used for livestock but not domestic purposes because of poor water quality (Bingham and Moore, 1975).

The terrace deposits in this area are unconsolidated, lens-shaped beds of sand, silt, clay, and gravel that were deposited by the North Canadian River during previous episodes of flooding. These deposits are very permeable and are a major source of water in the study area where thick enough to sustain acceptable yields (Bingham and Moore, 1975). Wells that yield water from only the terrace deposits tend to be shallow and produce water with low mineral content (Bingham and Moore, 1975).

The Permian-aged geologic units in the Garber–Wellington aquifer are the Garber Sandstone and the Wellington Formation. The Garber Sandstone consists mostly of lenticular beds of orange-brown fine-grained sandstone interbedded with siltstone and mudstone and small amounts of conglomerate (Breit, 1998). The lithology of the underlying Wellington Formation is similar to that of the Garber Sandstone but is finer-grained, containing a higher percentage of siltstone and mudstone. The stratigraphic contact between these two rock units is gradational, and discerning where one begins and the other ends is difficult (Smith and others, 2009; Parkhurst and others, 1996). Both units were deposited in a fluvial-deltaic environment at the edge of a shallow sea that periodically covered this part of Oklahoma during the Permian Period (Christenson and Havens, 1998). In general, the lithologies of the Garber Sandstone and Wellington Formation are highly variable over short distances, horizontally and with depth. The bedrock units dip in a westerly direction at about 50 ft per mile (Christenson and Havens, 1998) and become shalier (finer-grained) in an easterly direction and with depth (Schlottmann and others, 1998).

Recent surficial geologic mapping of the area by the Oklahoma Geological Survey indicates an absence of terrace deposits in the study area compared to maps prepared by Bingham and Moore (1975) (T. Stanley, Oklahoma Geological Survey, written commun., 2011). Where the terrace deposits are absent, wells in this area are completed in an upper permeable zone, possibly a highly weathered layer or regolith, of the Wellington Formation that underlies and is adjacent to the North Canadian River valley. Additionally, the contact between the Garber Sandstone and Wellington Formation has been mapped farther west of the study area than in the original mapping by Bingham and Moore (1975) (T. Stanley, Oklahoma Geological Survey, written commun., 2011), which would preclude the assumption that some of the wells sampled for this study were producing water from the Garber Sandstone. The figures in this report use the original mapping by Bingham and Moore (1975), but there is potential for different interpretation.

The terrace deposits and underlying bedrock contain unconfined aquifers (water table) that are hydraulically connected but may contain discontinuous strata that are relatively impermeable and locally impede the flow of groundwater from the surface and through the aquifer (Christenson and Havens, 1998). Groundwater in the study area flows in a southwesterly direction toward the North Canadian River (Mashburn and Magers, 2011).

The lack of completion logs for wells sampled for this study hindered the ability to identify depths of rock strata that are a source of trace elements or radionuclides in groundwater. Shallow wells generally produce relatively "young" water; however, the opposite is not always true, as deep wells do not always exclusively produce "older" water from deeper parts of an aquifer. Well casing may be open to multiple zones in an aquifer to maximize production from deep wells. As a result, water produced from deep wells can exhibit chemical characteristics of water from shallow, intermediate, and deep zones in the aquifer or mixtures from two or more zones. Drilling contractors installing private wells in the Garber–Wellington aquifer (and other aquifers) decide how a well should be constructed by examination of drill cuttings, depth to water, and other onsite information (Driscoll, 1986). After determining depths of the most favorable zones for producing water, well casing is inserted into the borehole with screened openings intersecting these layers. Loose sand is then used to fill the void space between the borehole and the well casing from the bottom of the hole to about 10–20

ft below land surface (Oklahoma Water Resources Board, 2011). Examination of available well completion logs in the surrounding area indicates that most wells have screened casing through at least two zones and sometimes as many as four (Oklahoma Water Resources Board, 2011). Local wells tend to have screened casing through a permeable layer near the upper portion of the borehole and a permeable layer at the bottom (Oklahoma Water Resources Board, 2011). For some wells, the upper portion is screened in the terrace deposits, and the bottom is screened in a sandstone layer in the underlying Wellington Formation (Oklahoma Water Resources Board, 2011). As a result of these factors, each well completion is unique, depending on the sequence of the aquifer rocks, distribution of the water-bearing zones, and judgment of the well driller.

Previous Studies in Oklahoma

For Permian- and Pennsylvanian-aged sandstones in Oklahoma, Al-Shaieb and others (1977) reported that uranium and radioactivity were associated with sandstones that had (1) a higher than average feldspar content, (2) oil production or seepage, (3) high concentrations of organic material, or (4) associations with evaporitic depositional environments. Christy and Pope (2009) reported that in Logan County the highest uranium concentrations were measured in water samples collected from wells greater than 295 ft in depth, concluding that the highest uranium concentrations were probably caused by deep groundwater interaction with uraniferous micas and clays in the Wellington Formation and the underlying Chase, Council Grove, and Admire Groups.

The most comprehensive study of trace elements and radioactivity in the Central Oklahoma aquifer was done by the USGS National Water Quality Assessment Program (NAWQA) from 1987 to 1992 (Christenson and Havens, 1998). That study examined rock cores, analyzed water-quality samples, and simulated water movement through the aquifer to describe the location, nature, and causes of selected water-quality problems (Christenson and Havens, 1998). With respect to water quality, the NAWQA study focused on arsenic, chromium, selenium, and uranium because these trace elements were the most likely to exceed public drinking-water MCLs in the aquifer and to affect water suppliers and consumers. These trace elements, which commonly form oxyanions (negatively charged molecules containing a metal atom and oxygen atoms), tend to be soluble in oxic (oxygenated) and alkaline groundwaters. Groundwaters of this type are commonly found in the United States west of the Mississippi River (Hodge and others, 1998; DeSimone, 2009; Ayotte and others, 2011).

Hematite, a mineral composed of iron and oxygen, coats sand grains, giving the Permian-aged sediments in Oklahoma a characteristic red color. The NAWQA study showed that the hematite coatings, other iron oxides, and clay minerals in mudstone are significant sources of oxidized arsenic, chromium, selenium, and uranium in the Central Oklahoma aquifer (Schlottmann and others, 1998). Concentrations of these trace elements in the aquifer rocks increase as rock particle sizes decrease, with the highest trace-element concentrations being measured in fine-grained sediment (clay-rich mudstone) (Schlottmann and others, 1998; Gromadzki, 2004). Negatively charged oxyanions tend to be strongly adsorbed to positively charged rims of hematite, other iron oxides, and clay minerals.

For the oxygenated groundwater of the Central Oklahoma aquifer, Parkhurst and others (1996) reported that concentrations of arsenic, chromium, selenium, and uranium are controlled by several factors, including concentrations of those elements in the aquifer rocks, water pH, water movement, and for uranium, the alkalinity of the water. The pH of the water in the Garber–Wellington aquifer at depth is controlled by the geochemical process of cation exchange. Cation exchange occurs primarily between sodium ions in fine-grained clay/mudstone layers and calcium and magnesium ions in groundwater that originate from the dissolution of dolomite in conglomerates in this aquifer. In the deeper parts of the aquifer, where the water is older, the exchange of sodium for calcium and magnesium in aquifer materials increases dissolution of dolomite in water, which increases pH, and, in sequence causes oxyanions of arsenic, chromium, selenium, and uranium to desorb from mineral surfaces (if present) and become mobile in water (Parkhurst and others, 1996, p. C64). Arsenic becomes mobile in groundwater at a pH of about 8.5 in the Garber–Wellington aquifer (Smith and Christenson, 2005; Becker and others, 2010); high chromium concentrations (50–100 µg/L) were measured at pH values equal to 8.3 or above, and selenium and uranium were measured at higher concentrations at pH values above 8.0 (Schlottmann and others, 1998). In addition to having high pH, groundwater that has undergone cation exchange characteristically is enriched with sodium. Calcium, magnesium, and bicarbonate are the dominant ions in groundwater at depths less than 90 ft (Christenson and Havens, 1998, fig. 13), but where the cation exchange process is occurring, generally in deeper parts of the aquifer, sodium-bicarbonate water type is predominant.

Methods

The initial phase of the study was designed to provide a reconnaissance assessment of groundwater pH and specific conductance in the study area. Values of pH and specific conductance were measured for water in the field from 59 private wells throughout the study area (fig. 3). Wells in the study area were selected on the basis of well access, knowledge of well depth, and the ability to access untreated water near the wellhead. Twenty of the 59 wells were subsequently selected and sampled (fig. 3) in summer and fall 2011. An effort was made to sample wells over a range of well depths and pH values and to sample wells that yielded water solely from each aquifer and from both aquifers concurrently; however, the lack of well-completion logs in the area hindered selecting wells on the basis of aquifer information.

Groundwater samples were analyzed for filtered concentrations of major ions, trace elements (including arsenic, chromium, and selenium), uranium, radium-226 and radium-228, and gross alpha- and gross beta-particle activities (table 1, apps. 1, 2, and 3). The concentration of radon-222 gas was analyzed in unfiltered water. In addition to alkalinity, the water properties temperature, dissolved oxygen concentration, pH, and specific conductance were measured in the field. Well w28 was sampled by the USGS in December 2008 for major ions and trace elements including uranium but was not sampled for other radionuclides. This well was destroyed prior to this study but is shown on figure 3 and the water-quality data are provided in appendixes 1, 2, and 3.

Uranium is a trace element that occurs naturally in soils, rocks, and water at trace concentrations but also is a radionuclide because all uranium isotopes undergo radioactive decay. In this report, uranium is grouped and described as a radionuclide and is reported in micrograms per liter. Concentrations of the other radionuclides analyzed for this study—radium-226, radium-228, and radon-222 gas—and gross alpha- and gross beta-particle activities are reported in picocuries per liter (table 1), with 1 pCi representing 2.2 radioactive disintegrations per minute.

Water-quality data were analyzed by determining water type from the predominant ions and making graphs to examine relations between constituents and common trends in water quality. In regards to water type, cations and anions were considered predominant when composing 50 percent or more of the total ion concentration expressed in milliequivalents per liter. Ions were considered to be secondary when composing between 25 and 49 percent of the total cation or anion concentration (Back, 1966).

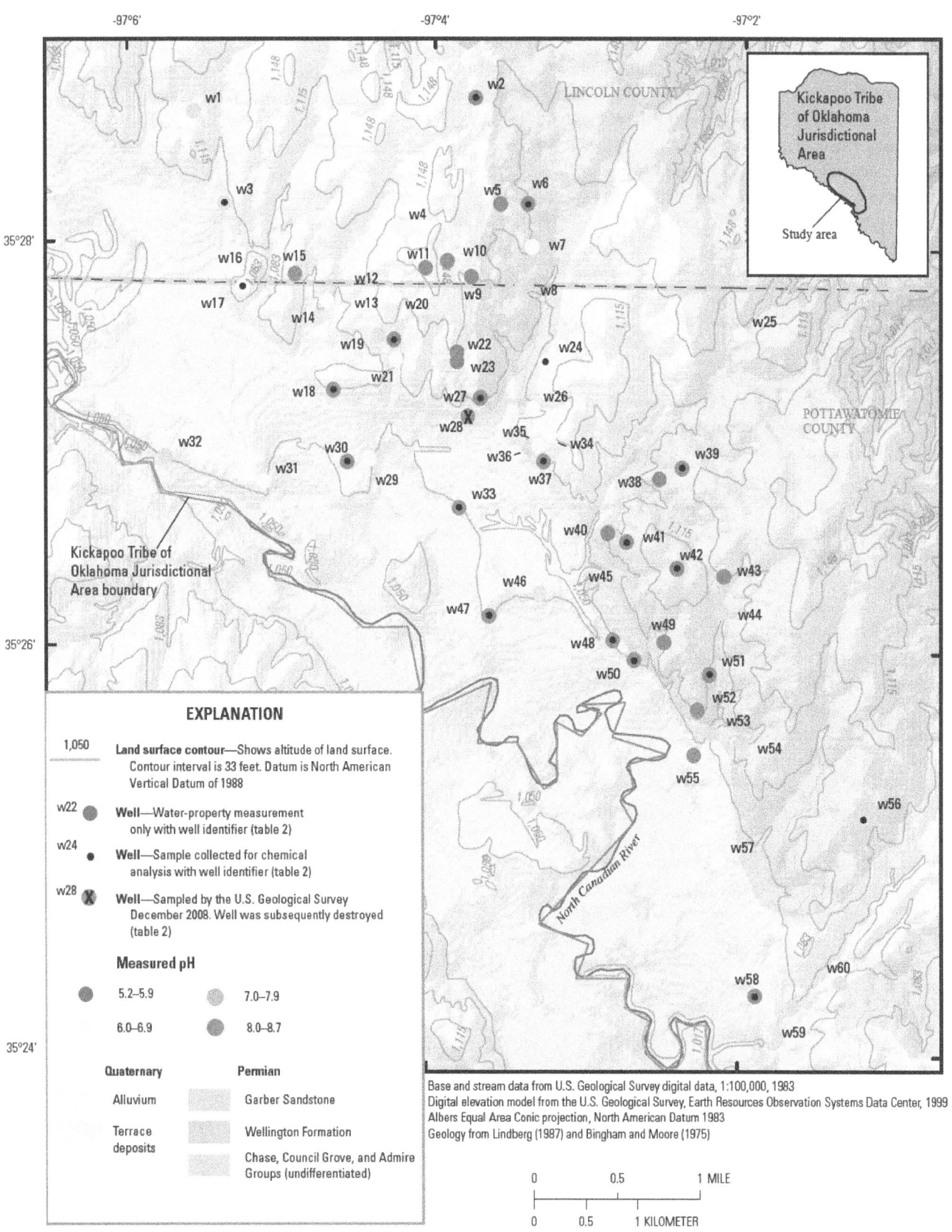

Figure 3.   Land surface contours, well locations, and pH measurements of water from private wells in part of the Kickapoo Tribe of Oklahoma Jurisdictional Area, central Oklahoma, 2011.

Table 1. Maximum Contaminant Levels, method references, and highest minimum laboratory reporting levels of water properties, major ions, trace elements, and radionuclides measured in water samples collected from 20 private wells in part of the Kickapoo Tribe of Oklahoma Jurisdictional Area, central Oklahoma, 2011.

[All concentrations are of filtered water unless otherwise noted; mg/L, milligrams per liter; —, not applicable; µS/cm, microsiemens per centimeter at 25 degrees Celsius; C, degree Celsius; $CaCO_3$, calcium carbonate; µg/L, micrograms per liter; pCi/L, picocuries per liter]

Water properties and chemical constituents (units)	Maximum Contaminant Level[1]	Secondary Maximum Contaminant Level[1]	Method references	Highest minimum laboratory reporting level
Water properties				
Dissolved oxygen, field (mg/L)	—	—	Wilde and Radtke (1998)	0.1
pH, field (standard units)	—	6.5–8.5	Wilde and Radtke (1998)	0.1 standard units
Specific conductance, field (µS/cm at 25°C)	—	—	Wilde and Radtke (1998)	3 significant digits
Water temperature, field (°C)	—	—	Wilde and Radtke (1998)	0.5
Major ions				
Calcium (mg/L)	—	—	Fishman (1993)	0.02
Magnesium (mg/L)	—	—	Fishman (1993)	0.012
Potassium (mg/L)	—	—	Fishman and Friedman (1989)	0.06
Sodium (mg/L)	—	—	Fishman (1993)	0.12
Alkalinity, field (mg/L as $CaCO_3$)	—	—	Rounds and Wilde (2001)	3 significant digits
Bicarbonate, field (mg/L)	—	—	Rounds and Wilde (2001)	3 significant digits
Carbonate, field (mg/L)	—	—	Rounds and Wilde (2001)	3 significant digits
Chloride (mg/L)	—	250	Fishman and Friedman (1989)	0.12
Fluoride (mg/L)	4	2	Fishman and Friedman (1989)	0.08
Silica (mg/L)	—	—	Fishman (1993)	0.02
Sulfate (mg/L)	—	250	Fishman and Friedman (1989)	0.18
Dissolved solids, sum (mg/L)	—	500	—	3 significant digits
Trace elements				
Antimony (µg/L)	6	—	Garbarino and others (2006)	0.027
Arsenic (µg/L)	10	—	Garbarino and others (2006)	0.022
Barium (µg/L)	2,000	—	Garbarino and others (2006)	0.07
Beryllium (µg/L)	4	—	Garbarino and others (2006)	0.006
Boron (µg/L)	—	—	Garbarino and others (2006); Garbarino (1999)	3
Bromide (µg/L)	—	—	Fishman and Friedman (1989)	0.02
Cadmium (µg/L)	5	—	Garbarino and others (2006)	0.016
Chromium (µg/L)	100	—	Garbarino and others (2006)	0.06
Cobalt (µg/L)	—	—	Garbarino and others (2006)	0.02
Copper (µg/L)[2]	1,300	1,000	Garbarino and others (2006)	0.05
Iron (µg/L)	—	300	Fishman (1993)	3.2
Lead (µg/L)[2]	15	—	Garbarino and others (2006)	0.015
Lithium (µg/L)	—	—	Garbarino and others (2006); Garbarino (1999)	0.22

Table 1. Maximum Contaminant Levels, method references, and highest minimum laboratory reporting levels of water properties, major ions, trace elements, and radionuclides measured in water samples collected from 20 private wells in part of the Kickapoo Tribe of Oklahoma Jurisdictional Area, central Oklahoma, 2011.—Continued

[All concentrations are of filtered water unless otherwise noted; mg/L, milligrams per liter; —, not applicable; µS/cm, microsiemens per centimeter at 25 degrees Celsius; C, degree Celsius; $CaCO_3$, calcium carbonate; µg/L, micrograms per liter; pCi/L, picocuries per liter]

Water properties and chemical constituents (units)	Maximum Contaminant Level[1]	Secondary Maximum Contaminant Level[1]	Method references	Highest minimum laboratory reporting level
Manganese (µg/L)	—	50	Garbarino and others (2006)	0.13
Molybdenum (µg/L)	—	—	Garbarino and others (2006)	0.014
Nickel (µg/L)	—	—	Garbarino and others (2006)	0.09
Selenium (µg/L)	50	—	Garbarino and others (2006)	0.03
Silver (µg/L)	—	100	Garbarino and others (2006)	0.005
Strontium (µg/L)	—	—	Garbarino (1999); Garbarino and others (2006)	0.2
Thallium (µg/L)	0.5	—	Garbarino and others (2006)	0.01
Vanadium (µg/L)	—	—	Garbarino and others (2006)	0.08
Zinc (µg/L)	—	5,000	Garbarino and others (2006)	1.4
Radionuclides				
Uranium, natural (µg/L)	30	—	Garbarino and others (2006)	0.004
Gross alpha particles (pCi/L)	15	—	National Environmental Methods Index (1980)	3
Gross beta particles (pCi/L)	4 millirems per year[3]	—	National Environmental Methods Index (1980)	4
Radium-226 (pCi/L)	5 (combined)	—	U.S. Environmental Protection Agency (2008a)	0.1
Radium-228 (pCi/L)		—	U.S. Environmental Protection Agency (1980)	1.0
Radon-222 (pCi/L) not filtered	Not finalized[4]	—	American Society of Testing and Materials International (2010)	20

[1]U.S. Environmental Protection Agency (2009a).

[2]Copper and lead are regulated by a treatment technique that requires systems to control the corrosiveness of the water. If more than 10 percent of tap water samples exceed the Maximum Contaminant Level, water systems must take corrective steps. For copper, the action level is 1,300 µg/L, and for lead it is 15 µg/L (U.S. Environmental Protection Agency, 2009a).

[3]The Maximum Contaminant Level is a concentration of beta particle activity from radionuclides in drinking water that produces an annual radiation dose of 4 millirems per year (U.S. Environmental Protection Agency, 2000).

[4]The Maximum Contaminant Level for radon in public water-supply systems will be 4,000 pCi/L if multimedia mitigation programs are established to address radon in indoor air. A second option is a Maximum Contaminant Level of 300 pCi/L for systems that choose not to develop multimedia mitigation programs (U.S. Environmental Protection Agency, 2012b).

Field Methods

Values of pH and specific conductance were measured for groundwater from 59 private wells (fig. 3) that ranged in depth from 60 to 320 ft. Well depth was determined (if possible) by lowering a steel tape in the well casing, by locating the well completion log, or from discussion with the well owner. Total depths of 10 wells were unknown. All wells had a water spigot either on or near the wellhead, enabling access to untreated water near the pressure tank. Measurements of pH and specific conductance were recorded while the well was pumping after

the measurements had stabilized using a YSI (YSI Inc., Yellow Springs, Ohio) multi-probe meter with the sonde immersed in a 5-gallon bucket.

Sampling for each of the 20 wells selected for major-ion, trace-element, and radionuclide analysis consisted of purging, measuring water properties, collecting water-quality samples, and preserving samples. Each well was purged of at least three standing well-casing volumes of water before sampling. The water properties—specific conductance, pH, temperature, and dissolved oxygen—were measured every 5–7 minutes starting halfway through the purging process by using a flow-through chamber with a YSI multiprobe meter (Wilde and Radtke, 1998). The meter calibrations were performed every morning before use. The pH and specific conductance calibrations used standard solutions that bracketed the expected values. The highest minimum reporting levels for water properties measured in the field are listed in table 1. Samples were collected after water properties had stabilized following the period of time needed to purge the well. Criteria for stabilization were less than a 0.2-unit variation in pH, less than a 10-percent variation in specific conductance, and less than a 0.3-mg/L variation in dissolved oxygen concentration. Some well pumps were operating when the sampling team arrived. Those wells were sampled after verifying that the volume of water purged was sufficient and that the water properties had stabilized.

Equipment used to sample each well consisted of a plastic Y-connector with a polypropylene adaptor (connects Y-connector to water spigot), a garden hose, and an adaptor for plastic tubing from which samples were collected. Water was filtered for dissolved constituents, collected in bottles, and preserved in an enclosed portable sampling chamber to prevent wind-borne contamination. Water collected for analysis of major ions, trace elements, gross alpha- and gross beta-particle activities, radium-226, and radium-228 was filtered through 0.45-micrometer polysulfone capsule filters and collected in acid-rinsed polyethylene bottles that had been prerinsed with filtered sample water. All water-quality samples were collected and processed by using established USGS protocols described in U.S. Geological Survey (2006) and Wilde and others (2004). Water-quality samples analyzed for trace elements were preserved by acidification with nitric acid to a pH of 2 or less. Sample bottles for anion analysis were not acid rinsed, nor were those samples acidified. Alkalinity, bicarbonate, and carbonate concentrations of filtered water were measured in the field or within 4 hours of sample collection by using an inflection-point titration method described by Rounds and Wilde (2001).

Water samples collected for radon-222 analysis were collected as described by U.S. Geological Survey (1987). Sample bottles for major ions and trace elements and vials for radon-222 were delivered to and analyzed by the USGS National Water Quality Laboratory in Lakewood, Colorado. Sample vials for radon-222 were shipped overnight for delivery within 24 hours. Sample bottles for gross alpha- and gross beta-particle activities and radium-226 and radium-228 were shipped overnight for delivery within 24 hours to Eberline Services in Richmond, California.

Laboratory Analysis

Major anion concentrations were measured by ion chromatography (IC) and ion-selective electrode (ISE) (Fishman and Friedman, 1989). Major cation concentrations were measured by using inductively coupled plasma-atomic emission spectrometry (ICP-AES) (Fishman, 1993). Trace elements were measured by using collision/reaction cell inductively coupled plasma-mass spectrometry (cICP-MS) (Garbarino and others, 2006; Garbarino, 1999) and inductively coupled plasma-mass spectrometry (ICP-MS) (Fishman, 1993) (table 1).

Gross alpha- and gross beta-particle activities were measured through gas-flow proportional counting at 72 hours after collection and again a second time approximately 30 days after the initial count by using a modification of EPA method number 900.0 (National Environmental Methods Index, 1980). Radium-226 was analyzed by using the radon emanation technique (EPA method number 903.1; U.S. Environmental Protection Agency, 2008a). Radium-228 was analyzed by using coprecipitation and beta counting, a modification of EPA method number 904.0 (U.S. Environmental Protection Agency, 1980). Radon was analyzed by using liquid scintillation following the American Society of Testing and Materials International standard test method for radon in drinking water D5072-09 (American Society of Testing and Materials International, 2010).

Reporting Results of Analysis of Radionuclides

The USGS National Water Quality Laboratory reports unrounded values for all radionuclide concentrations and activities in picocuries per liter (except for uranium, which is reported in micrograms per liter) with the 1-sigma combined standard uncertainty (CSU) and sample-specific critical level (ssLc) needed to analyze the results (app. 3). The CSU is the statistical standard deviation of an individual radionuclide concentration and is a function of several variables that can cause variances in measurement, one of which is the counting error associated with the random nature of radioactive decay within any given brief instant of time, which imposes limitation on the precision of radioactive counting techniques. The CSU is provided with the radionuclide concentration and defines upper and lower concentrations for a confidence interval of the true concentration (McCurdy and others, 2008). The ssLc is the lowest concentration measured that shows a significant statistical difference from the instrument background noise or the analytical blank sample (McCurdy and others, 2008). The ssLc is used as the detection level for determining if the radionuclide is present in a water sample. If the radionuclide concentration is greater than the ssLc, the radionuclide is considered to be present in a sample. The laboratory also reports a calculated sample-specific minimum detectable concentration (MDC) before a sample is analyzed to select appropriate analytical methods and parameters for the radionuclide measurement (McCurdy and others, 2008). In two cases, the ssLc for the sample exceeded the MDC because of unexpected chemical or instrument interference. These occurrences are noted as "C" in the "Remark" column in appendix 3.

Quality-Assurance Procedures

Decontamination of sampling equipment was performed by using USGS standard methods (Wilde, 2004). Quality-control samples consisted of an equipment blank and two replicates. The equipment blank was collected to determine if samples were contaminated by the sampling equipment or bottles. The equipment blank was prepared at the USGS Oklahoma Water Science Center laboratory with trace-metal-free blank water and analyzed for major cations, anions, and trace elements. The equipment blank was not analyzed for radionuclides. The equipment blank analysis (app. 4) indicated no contamination from the sampling equipment or bottles for the analyzed constituents.

A replicate sample is an extra sample set collected after the environmental sample to measure the variability of field and laboratory procedures. A replicate was collected from wells producing water with a high pH (w33, 8.2) and a low pH (w41, 5.5). The analytical accuracy between the environmental and replicate samples was computed as the relative percent difference (RPD) of constituent concentrations by using the following equation:

$$RPD = [(C1 - C2) / ((C1 + C2)/2)] \times 100], \qquad (1)$$

where
      C1   is the higher of the two concentrations, and
      C2   is the lower of the two concentrations.

Large RPD values can result from low concentrations reported with few significant figures. For example, concentrations of 2 and 3 would give an RPD of 40 percent, whereas if the concentrations were reported with more significant figures, such as 2.4 and 2.6, the RPD would be 8 percent.

RPD values for the eight major ions (app. 4) ranged from 0 to 3.3 percent in the environmental and replicate samples from both wells (w33 and w41), except for fluoride, which was 40 percent in samples from w41. RPD values for the 22 trace elements and uranium (app. 4) ranged from 0 to 40 percent. For trace-element concentrations, RPD values were highest for bromide, iron, manganese, selenium, and uranium, ranging from 16.2 to 40 percent.

Replicate information for radionuclides is shown in appendix 5. For both sets (w33 and w41) of replicate-sample analyses, concentrations of radium-226, radium-228, and radon-222 were within the boundaries of the confidence interval of the respective CSUs. Results were similar for both sets of replicate sample analyses for gross beta- and gross alpha-particle activities except for the gross alpha-particle activity determined 72 hours after sample collection for w33; those results were outside the boundaries of the confidence interval of the respective CSUs by 0.1 pCi/L, a relatively minor amount compared to the measured magnitude

of the gross alpha-particle activity. There was notable background (laboratory blank) activity for w33 (app. 5). Mean RPDs for both replicate sets were 12.5 percent for radium-226 and was slightly higher for gross alpha-particle activity (about 19.5 percent for the three samples with detectable results). The magnitude of RPD for the concentrations of both replicate samples was lower than the relative magnitude of the associated CSUs. Measurements were repeatable within the bounds imposed by the moderate analytical precision obtainable for radioactive counting techniques.

Groundwater Quality

The chemical characteristics of groundwater in the Garber–Wellington aquifer change with time and with depth; older water generally is found in deeper parts of the aquifer (Parkhurst and others, 1996). Minerals in rocks dissolve and react with water along groundwater flow paths through an aquifer in a process referred to as chemical evolution (Plummer and Back, 1980; Hem, 1985). With residence time in an aquifer, the chemical quality of water generally changes by increases in dissolved solids, with constituent concentrations being a product of the chemical composition of the aquifer rocks, groundwater, and biogeochemical processes in the aquifer (Plummer and Back, 1980; Ayotte and others, 2011).

In the Garber–Wellington aquifer, in addition to an increase in dissolved solids, chemical evolution of groundwater is indicated by a change in water type and an increase of pH resulting from cation-exchange processes in the aquifer (Schlottmann and others, 1998). The resulting increase in pH mobilizes oxyanions of arsenic, chromium, selenium, uranium, and other trace elements in groundwater (Schlottmann and others, 1998). Studies of other aquifers across the United States indicate that groundwater having high pH and oxidizing conditions commonly is associated with elevated concentrations of trace elements (Ayotte and others, 2007), especially oxyanions, including arsenic (Robertson, 1989; Welch and others, 2000) and uranium (Ayotte and others, 2011; Jurgens and others, 2009).

pH Values

Field measurements of groundwater pH from the 59 wells measured during the initial reconnaissance ranged from an acidic 5.2 to an alkaline 8.7, with a median value of 6.9 (figs. 3 and 4, tables 2 and 3). The Secondary Maximum Contaminant Level (SMCL) for pH in drinking water ranges from 6.5 to 8.5 for effects not related to health (U.S. Environmental Protection Agency, 2012c). Twenty-three wells (39 percent) had pH values below 6.5, and two wells (3 percent) had pH values above 8.5 (fig. 4). Seventy historical pH measurements in the Kickapoo Tribe of Oklahoma Jurisdictional Area ranged from 6.3 to 8.9 with a median and average pH of 7.5 (fig. 5A

Table 2. Field measurements, well information, and water type of water collected from private wells in part of the Kickapoo Tribe of Oklahoma Jurisdictional Area, central Oklahoma, 2011.

[ID, identifier; USGS, U.S. Geological Survey; μS/cm, microsiemens per centimeter at 25 degrees Celsius; —, not available. Vertical datum is North American Vertical Datum of 1988; blue shading indicates samples of mixed cation water types; red shading indicates samples of sodium-bicarbonate water type]

Well and sample ID	USGS station ID	pH	Specific conductance (μS/cm)	Land surface altitude (feet above vertical datum)	Well depth (feet below land surface)	Water type	Sampled for analysis
w1	352839097053401	7.4	443	1,099	160		No.
w2	352844097034401	5.9	132	1,126	120	Mixed cation	Yes.
w3	352812097052101	7.4	346	1,070	143	Mixed cation	Yes.
w4	352805097041301	6	230	1,127	95	—	No.
w5	352813097033401	5.6	639	1,133	60	—	No.
w6	352813097032401	5.7	277	1,112	61	Mixed cation	Yes.
w7	352800097032201	6.5	252	1,107	120	—	No.
w8	352748097032601	7.5	316	1,112	—	—	No.
w9	352751097034501	5.2	247	1,115	100	—	No.
w10	352756097035501	5.2	125	1,135	125	—	No.
w11	352753097040301	5.3	213	1,161	110	—	No.
w12	352751097041401	6.2	152	1,132	132	—	No.
w13	352745097041501	6.2	224	1,131	—	—	No.
w14	352743097044101	6	218	1,085	110	—	No.
w15	352751097045401	8.2	1,020	1,102	255	—	No.
w16	352751097050801	7.2	366	1,100	103	—	No.
w17	352747097051401	6.6	285	1,100	180	Mixed cation	Yes.
w18	352717097043801	8.3	522	1,077	140	Sodium-bicarbonate	Yes.
w19	352733097042101	7.5	327	1,107	110	—	No.
w20	352732097041501	5.9	184	1,100	102	Mixed cation	Yes.
w21	352726097041401	6	223	1,084	—	—	No.
w22	352728097035001	5.6	143	1,109	110	—	No.
w23	352726097035001	5.7	153	1,101	—	—	No.
w24	352726097031601	6.5	373	1,092	125	Mixed cation	Yes.
w25	352734097020001	6.6	275	1,106	222	—	No.
w26	352712097032101	6.2	306	1,076	125	—	No.
w27	352714097034101	8.5	690	1,087	135	Sodium-bicarbonate	Yes.
w28[1]	352710097034601	8.2	443	1,072	220	Sodium-bicarbonate	Yes.
w29	352657097042401	6.8	563	1,051	80	—	No.
w30	352655097043201	8.7	870	1,052	82	Sodium-bicarbonate	Yes.
w31	352652097050501	6.9	810	1,060	120	—	No.
w32	352657097054401	7.5	506	1,061	80	—	No.
w33	352642097034801	8.2	632	1,150	108	Sodium-bicarbonate	Yes.
w34	352704097031401	6.2	261	1,110	150	—	No.
w35	352701097031901	6	424	1,090	—	—	No.
w36	352700097032201	7.4	500	1,076	—	—	No.
w37	352656097031601	8.4	687	1,082	320	Sodium-bicarbonate	Yes.
w38	352651097023201	5.6	181	1,090	120	—	No.
w39	352654097022301	8	443	1,101	160	Sodium-bicarbonate	Yes.
w40	352635097025201	5.7	251	1,082	150	—	No.

Table 2. Field measurements, well information, and water type of water collected from private wells in part of the Kickapoo Tribe of Oklahoma Jurisdictional Area, central Oklahoma, 2011.—Continued

[ID, identifier; USGS, U.S. Geological Survey; μS/cm, microsiemens per centimeter at 25 degrees Celsius; —, not available. Vertical datum is North American Vertical Datum of 1988; blue shading indicates samples of mixed cation water types; red shading indicates samples of sodium-bicarbonate water type]

Well and sample ID	USGS station ID	pH	Specific conductance (μS/cm)	Land surface altitude (feet above vertical datum)	Well depth (feet below land surface)	Water type	Sampled for analysis
w41	352632097024401	5.5	156	1,091	155	Mixed cation	Yes.
w42	352625097022401	5.6	146	1,138	172	Mixed cation	Yes.
w43	352622097020701	5.4	239	1,119	—	—	No.
w44	352612097020701	7.8	792	1,085	114	—	No.
w45	352620097024201	7.6	513	1,095	150	—	No.
w46	352617097031801	7.7	504	1,053	180	—	No.
w47	352609097033801	8	559	1,057	180	Sodium-bicarbonate	Yes.
w48	352603097024901	8.4	673	1,051	163	Sodium-bicarbonate	Yes.
w49	352603097022901	5.7	336	1,085	125	—	No.
w50	352557097024101	8.5	809	1,052	180	Sodium-bicarbonate	Yes.
w51	352553097021101	8.4	783	1,081	167	Sodium-bicarbonate	Yes.
w52	352543097021601	8.1	809	1,092	160	—	No.
w53	352536097021101	7.8	709	1,052	—	—	No.
w54	352533097015901	7.7	626	1,084	240	—	No.
w55	352529097021801	8.6	733	1,042	103	—	No.
w56	352510097011201	7.3	693	1,103	142	Sodium-bicarbonate	Yes.
w57	352458097020801	6.9	744	1,041	—	—	No.
w58	352417097015301	8.3	415	1,043	110	Sodium-bicarbonate	Yes.
w59	352415097013901	7.9	407	1,042	—	—	No.
w60	352422097012101	7.8	402	1,055	90	—	No.

[1]Well was sampled and water properties measured by the USGS in December 2008. This well was subsequently destroyed.

and table 3). The range of pH values measured for this study was substantially lower than historical pH values; groundwater from 23 wells sampled for this study had pH values below the historical low pH of 6.3. Acidic groundwater can be a problem for well owners because it can corrode metal plumbing and fixtures (National Environmental Education Foundation, 2011). To reduce the intake of shallow acidic water into wells, well drillers may extend and cement the outside well casing to a depth greater than the minimum required (Oklahoma Water Resources Board, 2011). For example, well w15, which produced water having a pH of 8.2, was constructed (as reported by the well owner) to exclude water from the terrace deposits by cementing the upper 40 ft of the borehole. The depths of zones with screened casing were unknown, but this well was probably completed in the Wellington Formation. The driller's log for well w17 shows it was also probably completed in the Wellington Formation and constructed to exclude shallow water by cementing the upper 80 ft of the borehole (Oklahoma Water Resources Board, 2011). This well was screened at 120–130 ft and 165–180 ft and yielded water with a pH of 6.6.

Field measurements of specific conductance, used as a surrogate for dissolved solids, ranged from 125 to 1,020 microsiemens per centimeter (μS/cm) with a median of 402 μS/cm in water from the 59 wells (table 3). Field measurements of pH and specific conductance show that values of both water properties tended to decrease in wells located at higher altitudes. Wells located at higher altitudes, on the terrace deposits and above the North Canadian River Valley, tended to produce water having lower pH and specific conductance values than wells at lower altitudes (figs. 6A and 6B, table 2). Wells at higher altitudes tend to be shallower and probably yield a greater contribution of water from the terrace deposits or shallow permeable zones in the Wellington Formation. An exception is well w30, which is relatively shallow, 82 ft, and has anomalously high pH (8.7) and specific conductance (870 μS/cm) values (fig. 3 and table 2). This well had been placed adjacent to (less than 10 ft from) a deeper well (300 ft deep) that had been previously rendered unusable. The high pH and specific conductance values probably reflect the contribution of deeper water from the adjacent borehole.

Table 3. Statistical summaries of pH, specific conductance, and concentrations of selected trace elements and radionuclides in water samples collected from 20 private wells for this study (2011) and in historical analyses in the Kickapoo Tribe of Oklahoma Jurisdictional Area, central Oklahoma.

[MCL, Maximum Contaminant Level; SMCL, Secondary Maximum Contaminant Level; mg/L, milligrams per liter; µg/L, micrograms per liter; pCi/L, picocuries per liter; µS/cm, microsiemens per centimeter; ft, foot; na, not applicable; —, not calculated because of censored values; ND, not detected]

Constituent	Number of wells	SMCL–MCL[1]	Number of wells exceeding SMCL–MCL[1]	Minimum	Percentiles			Maximum	Mean
					25	50 median	75		
Historical samples (1954–2008)[2,3]									
pH	70	6.5–8.5	2	6.3	7.0	7.5	7.9	8.9	7.5
Specific conductance (µS/cm)	72	na	na	100	558	755	945	7,320	931
Well depth (ft)	72	na	na	20	57	100	140	412	112
Arsenic (µg/L)	46	10	2	0.25	0.25	0.37	1	18	1.5
Uranium (µg/L)	43	30	6	0.01	0.2	1.8	15	1,500	50
Gross alpha-particle activity (pCi/L)	8	15	3	1.4	2.1	2.8	31	210	39
Radon-222	10	[4]300–4,000	2/1	<80	135	210	292	4,900	698
Sampled wells									
pH	20	6.5–8.5	6	5.5	6.3	8	8.4	8.7	7.4
Specific conductance (µS/cm)	20	na	na	132	283	482	688	870	484
Dissolved oxygen (mg/L)	19	na	na	0.8	3	3.9	5.9	8.9	4.2
Total dissolved solids (mg/L)	20	500	8	81	160	283	400	501	282
Arsenic (µg/L)	20	10	1	0.07	0.20	0.59	1.4	25	2.6
Chromium (µg/L)	20	100	0	<0.06	—	1.2	—	31	—
Selenium (µg/L)	20	50	1	<0.03	—	0.93	—	147	—
Uranium (µg/L)	20	30	5	0.02	0.20	1.54	33	383	48
Gross alpha-particle activity (72 hours) (pCi/L)	20	15	8	ND	1.9	5.3	54	370	54
Gross alpha-particle activity (30 days) (pCi/L)	20	15	8	ND	0.6	7.3	51	350	51
Gross beta-particle activity (72 hours) (pCi/L)	20	50	0	ND	0.6	1.3	2	8.6	1.8
Gross beta-particle activity (30 days) (pCi/L)	20	50	0	0.8	1.5	2.1	9.6	102	14
Radium (226 and 228 combined) (pCi/L)	20	5	0	0.03	0.12	0.45	0.70	1.7	0.52
Radon-222 (pCi/L)	20	[4]300–4,000	8/0	95	168	261	467	3,600	580

Table 3. Statistical summaries of pH, specific conductance, and concentrations of selected trace elements and radionuclides in water samples collected from 20 private wells for this study (2011) and in historical analyses in the Kickapoo Tribe of Oklahoma Jurisdictional Area, central Oklahoma.—Continued

[MCL, Maximum Contaminant Level; SMCL, Secondary Maximum Contaminant Level; mg/L, milligrams per liter; µg/L, micrograms per liter; pCi/L, picocuries per liter; µS/cm, microsiemens per centimeter; ft, foot, na, not applicable; —, not calculated because of censored values; ND, not detected]

Constituent	Number of wells	SMCL–MCL[1]	Number of wells exceeding SMCL–MCL[1]	Minimum	Percentiles 25	Percentiles 50 median	Percentiles 75	Maximum	Mean
Field measurements									
pH	59	6.5–8.5	25	5.2	6.0	6.9	7.9	8.7	6.9
Specific conductance (µS/cm)	59	na	na	125	243	402	636	1,020	438
Samples having sodium-bicarbonate water type									
pH	12	6.5–8.5	1	7.3	8.1	8.4	8.4	8.7	8.2
Specific conductance (µS/cm)	12	na	na	415	550	680	715	870	648
Dissolved oxygen (mg/L)	11	na	na	0.8	1.6	3.2	3.7	5.5	2.9
Total dissolved solids (mg/L)	12	500	1	248	326	394	423	501	380
Arsenic (µg/L)	12	10	1	0.15	0.56	0.82	5.25	24.7	4.2
Chromium (µg/L)	12	100	0	<0.06	—	4.6	—	31	—
Selenium (µg/L)	12	50	1	<0.03	—	8.6	—	147	—
Uranium (µg/L)	12	30	5	1.1	3.9	27	61	383	79
Gross alpha-particle activity (72 hours) (pCi/L)	12	15	8	5	10	42	97	370	88
Gross alpha-particle activity (30 days) (pCi/L)	12	15	8	4	11	41	89	350	85
Gross beta-particle activity (72 hours) (pCi/L)	12	50	0	ND	0.98	1.8	2.5	8.6	2.4
Gross beta-particle activity (30 days) (pCi/L)	12	50	0	1.8	3.2	8.1	18	102	23
Radium (226 and 228 combined) (pCi/L)	12	5	0	0.03	0.27	0.49	0.85	1.7	0.65
Radon-222 (pCi/L)	12	[4]300–4,000	8/0	143	242	395	1,085	3,600	843
Samples having mixed cation water type									
pH	8	6.5–8.5	5	5.5	5.7	5.9	6.5	7.4	6.1
Specific conductance (µS/cm)	8	na	na	132	153	230	300	373	237
Dissolved oxygen (mg/L)	8	na	na	2.9	5.9	5.9	6.5	8.9	6.1
Total dissolved solids (mg/L)	8	500	0	81	90	134	169	198	135
Arsenic (µg/L)	8	10	0	0.07	0.14	0.19	0.33	0.63	0.27
Chromium (µg/L)	8	100	0	0.12	0.45	0.75	1.0	2.7	0.91

Table 3. Statistical summaries of pH, specific conductance, and concentrations of selected trace elements and radionuclides in water samples collected from 20 private wells for this study (2011) and in historical analyses in the Kickapoo Tribe of Oklahoma Jurisdictional Area, central Oklahoma.—Continued

[MCL, Maximum Contaminant Level; SMCL, Secondary Maximum Contaminant Level; mg/L, milligrams per liter; µg/L, micrograms per liter; pCi/L, picocuries per liter; µS/cm, microsiemens per centimeter; ft, foot; na, not applicable; —, not calculated because of censored values; ND, not detected]

Constituent	Number of wells	SMCL–MCL[1]	Number of wells exceeding SMCL–MCL[1]	Minimum	Percentiles			Maximum	Mean
					25	50 median	75		
Samples having mixed cation water type—Continued									
Selenium (µg/L)	8	50	0	0.15	0.21	0.26	0.54	1.2	0.47
Uranium (µg/L)	8	30	0	0.02	0.07	0.14	0.22	0.47	0.16
Gross alpha-particle activity (72 hours) (pCi/L)	8	15	0	ND	0.87	1.2	2.5	5.2	1.9
Gross alpha-particle activity (30 days) (pCi/L)	8	15	0	ND	ND	ND	0.82	1	0.34
Gross beta-particle activity (72 hours) (pCi/L)	8	50	0	ND	0.6	1.2	1.4	1.7	0.98
Gross beta-particle activity (30 days) (pCi/L)	8	50	0	0.8	0.9	1.2	1.7	1.7	1.3
Radium (226 and 228 combined) (pCi/L)	8	5	0	0.04	0.12	0.29	0.52	0.70	0.32
Radon-222 (pCi/L)	8	[4]300–4,000	0/0	95	141	168	256	278	187

[1]U.S. Environmental Protection Agency, 2009a.

[2]Includes well w28 sampled in December 2008 by the U.S. Geological Survey. This well was subsequently destroyed.

[3]Historical water-quality data were retrieved from the U.S. Geological Survey National Water Information System database on December 1, 2011 (U.S. Geological Survey, 2012).

[4]The Maximum Contaminant Level for radon in public water-supply systems will be 4,000 pCi/L if multimedia mitigation programs are established to address radon in indoor air. A second option is a Maximum Contaminant Level of 300 pCi/L for systems that choose not to develop multimedia mitigation programs (U.S. Environmental Protection Agency, 2012b).

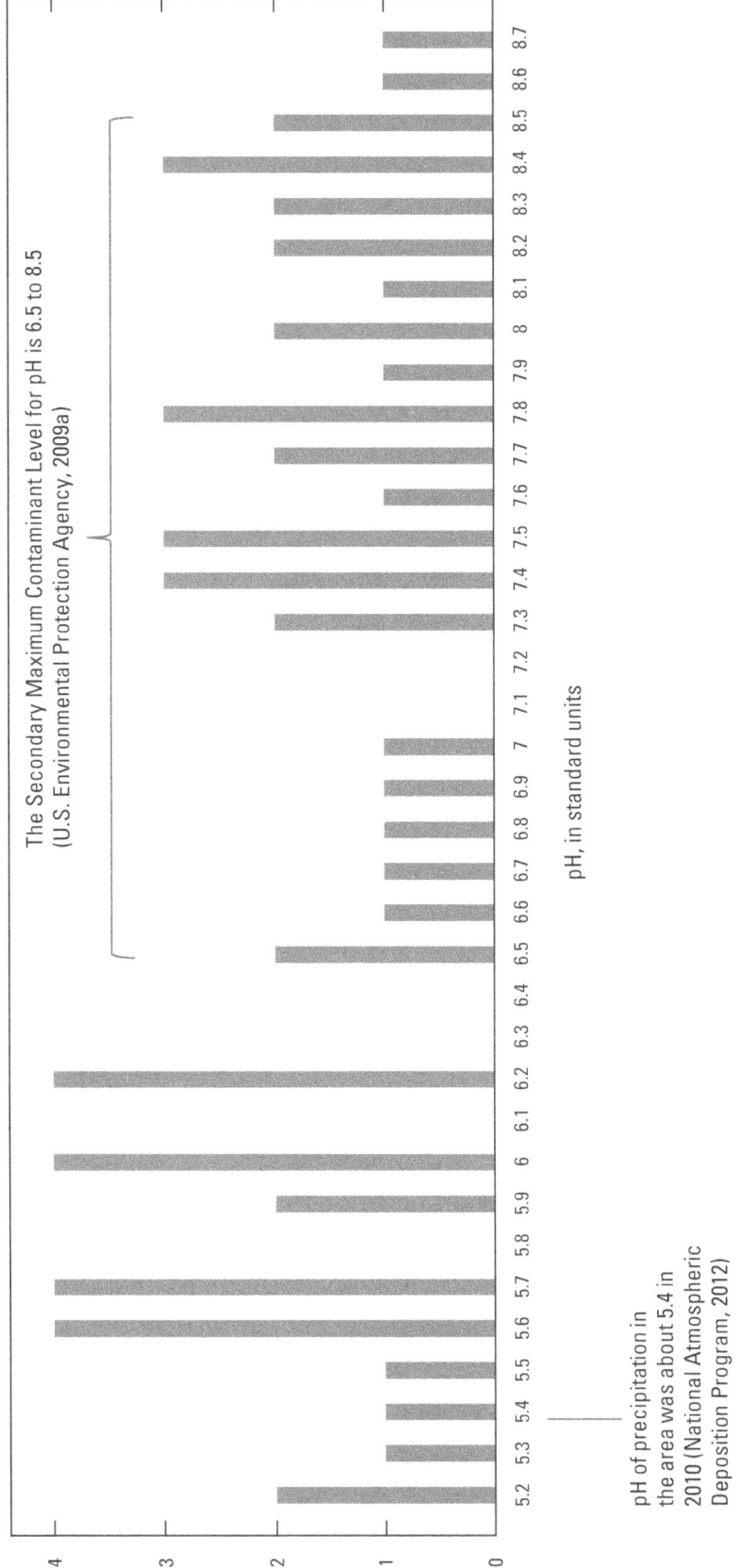

Figure 4.   Number of wells and pH values measured in water collected from 59 private wells in part of the Kickapoo Tribe of Oklahoma Jurisdictional Area, central Oklahoma, 2011.

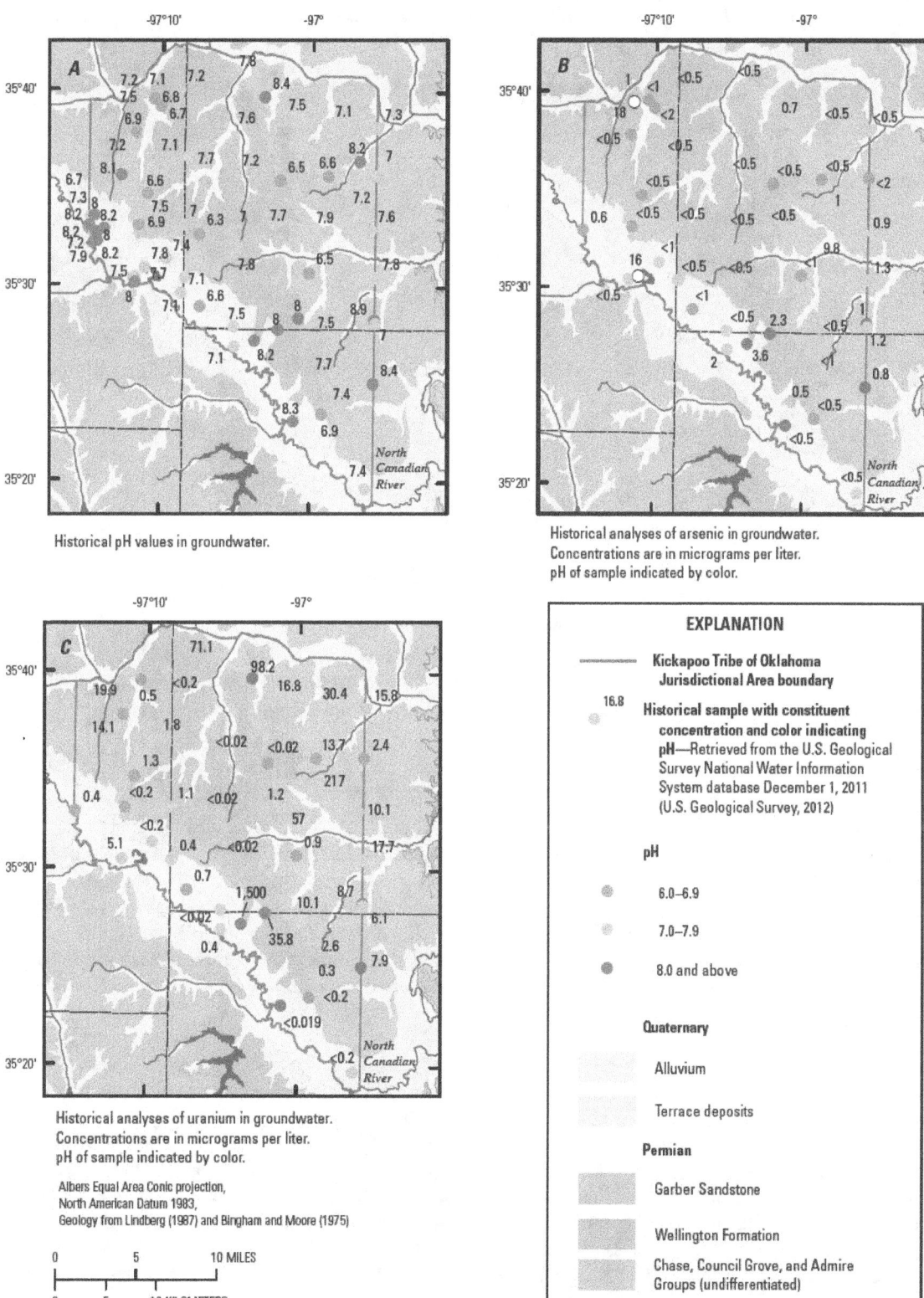

Figure 5. Historical water-quality data from wells in the Kickapoo Tribe of Oklahoma Jurisdictional Area, central Oklahoma.
A, Historical pH values in groundwater (1954–2008). *B*, Historical pH and analyses of arsenic in groundwater (concentrations in micrograms per liter) (1977–2008). *C*, Historical pH and analyses of uranium in groundwater (concentrations in micrograms per liter) (1977–2008).

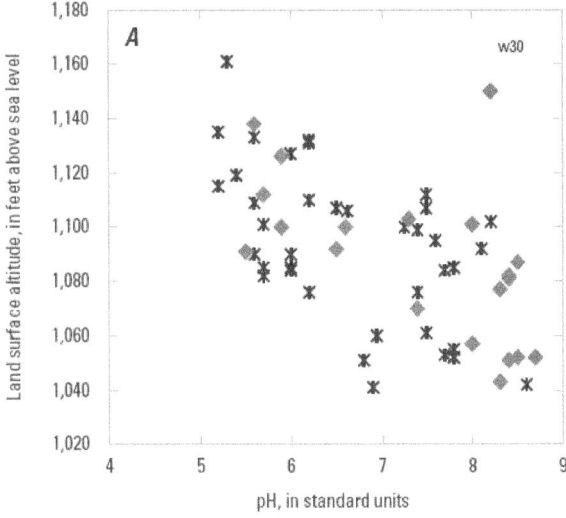 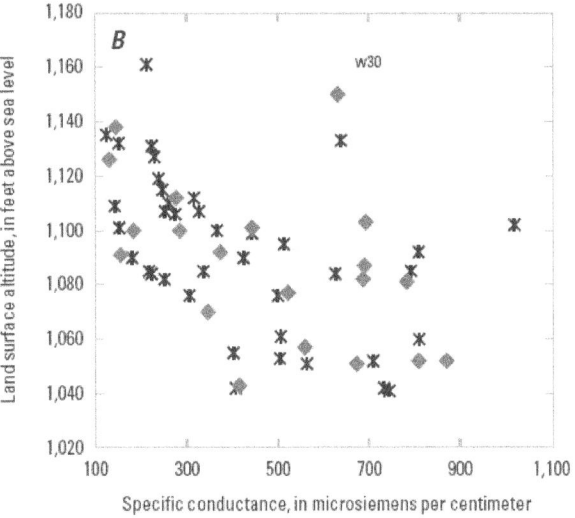

EXPLANATION

✖   **Well sample**—Water-property measurements only

◆   **Well sample**—Sodium-bicarbonate water type

◆   **Well sample**—Mixed cation water type

Figure 6.   Relation of land surface altitude of wells to measurements of *A*, pH and *B*, specific conductance of water collected from 59 private wells in part of the Kickapoo Tribe of Oklahoma Jurisdictional Area, central Oklahoma, 2011.

The primary source of recharge to the terrace deposits (and Garber–Wellington aquifer) is precipitation (Parkhurst and others, 1996), which has a relatively low pH and low mineral content. The average pH of precipitation in this area of the State in 2010 was 5.4 (National Atmospheric Deposition Program, 2012).

Major Ions and Water Types

Groundwater samples from the 20 wells sampled for chemical analyses for the study varied from a sodium-bicarbonate water type, characteristic of older water having high pH values, to water having multiple ions characteristic of younger water having lower pH values. As shown on the Piper diagram (Piper, 1944) in figure 7, sodium and bicarbonate were the dominant ions in samples from 12 of the 20 wells (table 2 and fig. 7). These 12 samples had high pH values ranging from 7.3 to 8.7 with a median pH of 8.4 and concentrations of dissolved solids ranging from 248 to 501 mg/L with a median value of 394 mg/L (specific conductance ranged from 415 to 870 µS/cm with median of 680 µS/cm) (table 3). One sample, w30, was the only sample of 20 collected that exceeded the SMCL for dissolved solids, having a concentration of 501 mg/L (app. 1). Both pH and specific conductance of the water from these 12 wells were

higher than for water from the other 8 wells, which were characterized as a mixed cation water type (table 3). Those 12 wells were located mostly in the river valley at lower altitudes and probably yield water from deeper zones of the Garber–Wellington aquifer. The high pH and sodium-bicarbonate water type of these samples are indications of the cation-exchange process occurring in the aquifer.

Samples from 5 of the 20 wells (w2, w3, w17, w20, and w24) were a mixed cation-bicarbonate water type having a mixture of calcium, sodium, and magnesium cations in various ratios and bicarbonate as the dominant anion (table 2 and fig. 7). Samples from 3 of the 20 wells, w6, w41, and w42, also were a mixed cation water type having a mixture of calcium, sodium, and magnesium cations in different ratios but with chloride dominant or a mixture of chloride, bicarbonate, and sulfate anions (table 2 and fig. 7). These three samples had the lowest dissolved solids concentrations (157, 89, and 81 mg/L) and the lowest pH values (5.7, 5.5, and 5.6) and show the least evidence of cation exchange. Water from these wells is probably relatively young water, as indicated by the low pH values and dissolved solids concentrations. Values of pH of the eight samples having mixed cation water types ranged from 5.5 to 7.4, with a median of 5.9, and dissolved solids concentrations ranging from 81 to 198 mg/L, with a median 134 mg/L (table 3).

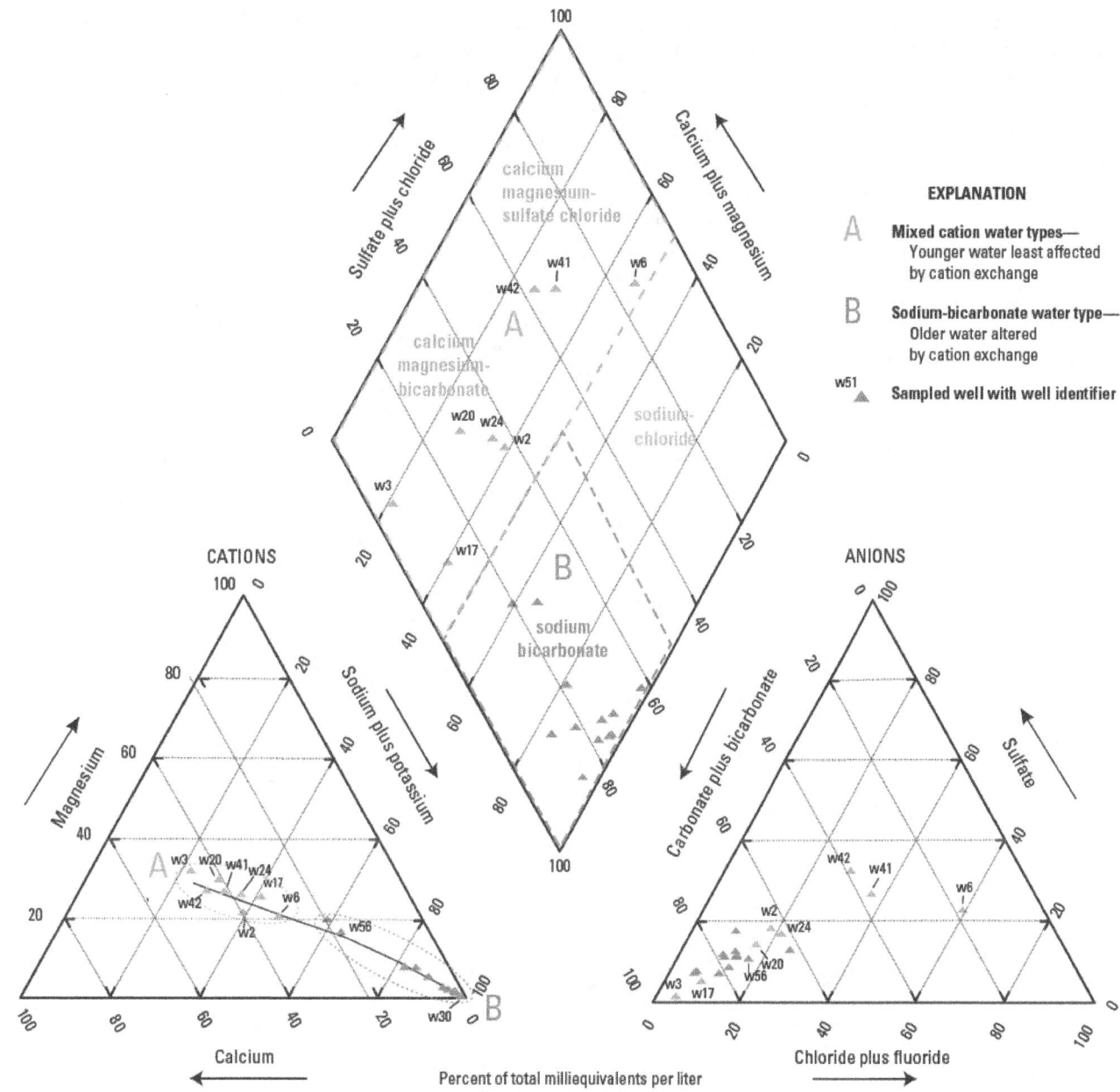

The letter A on the cation triangle of the Piper diagram labels the groups of samples composed of mixed cations. The letter B labels the samples of older water classified as sodium-bicarbonate water type. The line connecting the end members, w3 and w30, symbolizes all possible mixtures of water represented by the two end members. Water from a well producing from multiple zones in the aquifer will fall somewhere near this line, depending on the percentage of contribution from each zone. This line also represents the process of cation exchange and shows the evolution of cation replacement of calcium and magnesium by sodium.

Figure 7.   Water types and percentages of cations and anions composing water samples collected from 20 private wells in part of the Kickapoo Tribe of Oklahoma Jurisdictional Area, central Oklahoma, 2011.

The letter "A" on the cation triangle of the Piper diagram labels the groups of samples composed of mixed cations (fig. 7). Sample w3 is shown as the least altered sample from cation exchange and represents the end member of this group. The letter "B" labels the samples of older water classified as sodium-bicarbonate water type. Sample w30 is shown to be the most altered from cation exchange and represents the end member of this water type group. The line connecting the end members, w3 and w30, symbolizes all possible mixtures of water represented by the two end members. Water from a well producing from multiple zones in the aquifer will fall somewhere near this line, depending on the percentage of contribution from each zone. This line also represents the process of cation exchange and shows the evolution of cation replacement of calcium and magnesium by sodium.

Anion concentrations are not directly affected by cation exchange and do not plot in distinctive groups on the anion triangle similar to the way in which cations do; however, when the percentages of anions are plotted with cations on the central diamond, samples plot in quadrants of the diamond, revealing the general composition of water and the water types (fig. 7). Samples dominated by sodium and bicarbonate ions plot in the lower quadrant, whereas the two sample groups composed of mixed cations plot in two quadrants classified as calcium magnesium-bicarbonate and calcium magnesium-sulfate chloride water types.

Trace Elements

Water samples collected from the 20 wells for chemical analyses were analyzed for 22 trace elements, of which 10 have MCLs and 5 have SMCLs (table 1) (U.S. Environmental Protection Agency, 2009a). Samples from two wells had concentrations exceeding an MCL: the sample from well w39 had an arsenic concentration of 24.7 µg/L (exceeding the MCL of 10 µg/L), and the sample from well w37 had a selenium concentration of 147 µg/L (exceeding the MCL of 50 µg/L) (figs. 8A and 8C, app. 2). Both of these samples were a sodium-bicarbonate water type and had high pH values, 8.0 and 8.4,

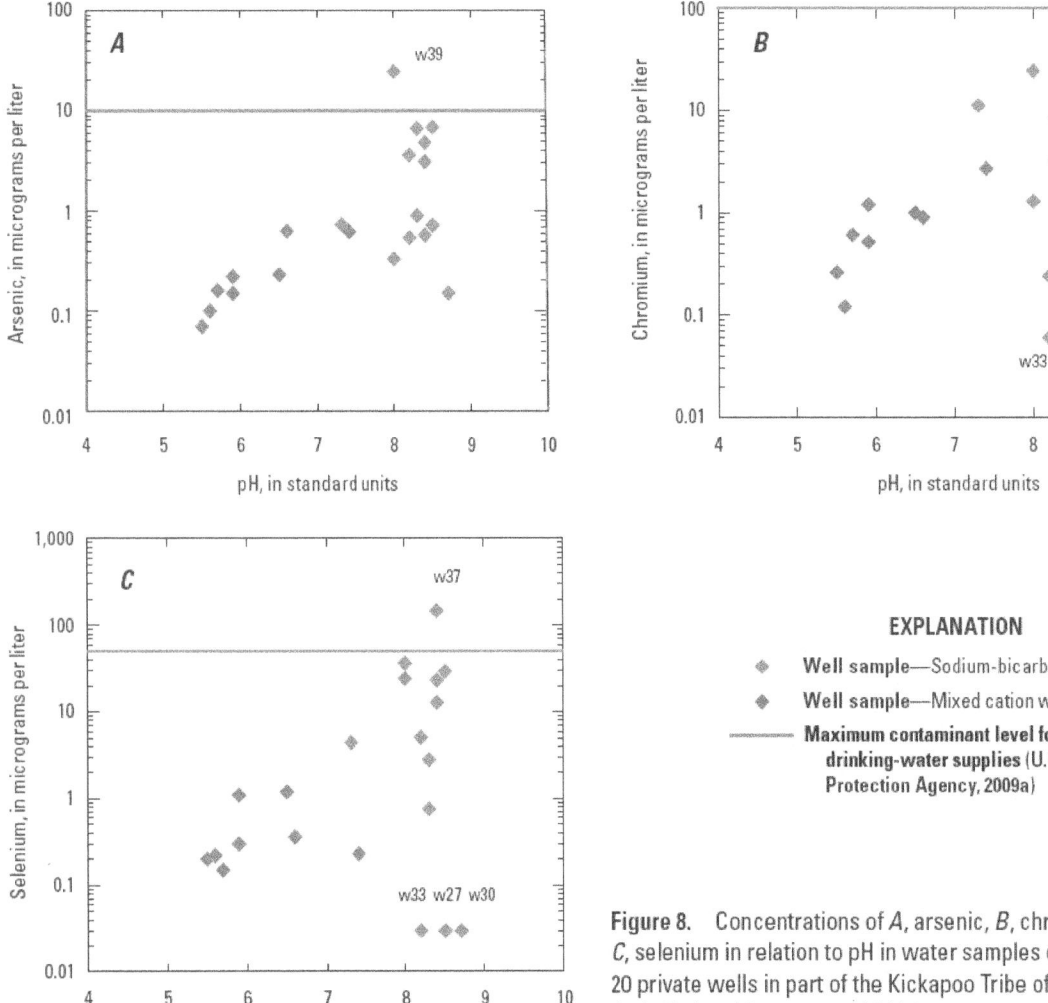

EXPLANATION

- ⬧ Well sample—Sodium-bicarbonate water type
- ⬧ Well sample—Mixed cation water type
- ─── Maximum contaminant level for public drinking-water supplies (U.S. Environmental Protection Agency, 2009a)

Figure 8.   Concentrations of A, arsenic, B, chromium, and C, selenium in relation to pH in water samples collected from 20 private wells in part of the Kickapoo Tribe of Oklahoma Jurisdictional Area, central Oklahoma, 2011.

respectively. All other trace-element concentrations were below MCLs and SMCLs for public drinking-water supplies (app. 2).

A review of historical analyses indicated there were few trace-element concentrations in water samples that exceeded MCLs. Only 2 of 46 historical analyses for arsenic (16 and 18 µg/L) exceeded the MCL; the pH of those two samples is unknown. There were no historical samples exceeding the MCLs for chromium concentration in 42 samples or selenium concentration in 30 samples. Because of relatively high laboratory reporting levels, most historical trace-element concentrations were censored and were not useful for determining relations between trace-element concentrations and pH values.

Concentrations of many trace elements in the collected water samples show a relation to pH. Arsenic, chromium, and selenium concentrations had a positive relation to pH, with the highest concentrations occurring in samples having pH values above 8.0 (figs. 8A, 8B, and 8C). Three samples, however—w27, w30, and w33, with pH values above 8.0—had no detectable concentrations of chromium and selenium (figs. 8B and 8C). This absence of detection may indicate the lack of these trace elements in the aquifer matrix along groundwater-flow paths to these wells or that geochemical controls other than pH are affecting trace-element mobility in groundwater.

In water samples collected for this study, other trace elements having increasing concentrations with increasing pH were boron, molybdenum, and vanadium (fig. 9). In the form of oxyanions, these trace elements can compete for sorption sites on mineral surfaces and can be released into groundwater during the competitive sorption and cation-exchange process (Hem, 1985; Robertson, 1989; Ayotte and others, 2011). Similar to selenium, vanadium had no detectable concentrations in samples from w33, w27, and w30 (fig. 9). Concentrations of barium, lithium, and strontium substantially decreased at pH values near and above 8.0. Barium and strontium are chemically similar to divalent calcium and magnesium and may be preferentially removed from the groundwater during cation exchange. Lithium is chemically similar to monovalent sodium and may be incorporated into clay minerals during cation exchange. Beryllium (not shown) and nickel had negative relations to pH, with concentrations of those trace elements being highest at pH values less than 6.0 (fig. 9); these constituents tend to undergo hydrolysis but in acidic water are commonly present as free cations that are sparingly soluble (Ayotte and others, 2011).

Radionuclides

Each radionuclide analyzed for this study has unique chemical properties and behaves differently in groundwater depending on the geochemical conditions in an aquifer (Zapecza and Szabo, 1988). Elevated concentrations of uranium and radon in groundwater have been reported to be associated with oxygen-rich groundwater having relatively high alkalinities and pH values (Zapecza and Szabo, 1988; Ayotte and others, 2007; Jurgens and others, 2009) similar to

conditions in the Garber–Wellington aquifer. Radium-226, a decay product of uranium, and radium-228, a decay product of thorium, however, are most commonly associated with oxygen-depleted (anoxic) groundwater, low pH values, and high concentrations of dissolved solids and chlorides (Szabo and others, 2012). Radon-222 may be present in elevated concentrations under a wide range of chemical conditions in groundwater, but concentrations of that isotope are dependent in part on the presence of the immediate parent isotope, radium-226, in aquifer materials or in groundwater.

Concentrations of uranium, gross alpha-particle activity, and radon-222 were highest in the 12 samples having a sodium-bicarbonate water type with pH values of 8.0 or more (figs. 10A, 10B, and 10C). Uranium concentrations ranged from 0.02 to 383 µg/L with a median of 1.54 µg/L in the 20 samples (table 3), with 5 samples exceeding the MCL of 30 µg/L. Samples from these wells (w18, w27, w37, w39, and w51) had uranium concentrations ranging from 48.3 to 383 µg/L, with pH values ranging from 8.0 to 8.5 (fig. 10A). Samples from three wells (w30, w33, and w58), also having high pH and a sodium-bicarbonate water type, had uranium concentrations less than 2 µg/L and were the only samples of this water type that did not exceed the MCL for gross alpha-particle activity (fig. 10B). Uranium concentrations in the eight samples having mixed cation water types were considerably lower, ranging from 0.02 to 0.47 µg/L (fig. 10A). These waters had lower pH and bicarbonate concentrations than the sodium-bicarbonate water type. Seven of 43 historical analyses (including well w28) in the tribal jurisdictional area had uranium concentrations exceeding the MCL of 30 µg/L, ranging from 30.4 to 1,500 µg/L, and had pH values ranging from 7.1 to 8.4 (figs. 5C and 10A). Both uranium concentrations and gross alpha-particle activity had positive relations to pH, with the highest concentrations and activities being associated with pH values of 8.0 and higher (figs. 10A and 10B).

In oxidizing groundwater such as that in the Garber–Wellington aquifer, bicarbonate enhances desorption of uranium from mineral surfaces in aquifer rocks (Hsi and Langmuir, 1985; Curtis and others, 2004), forming uranium-carbonate complexes in groundwater (Langmuir, 1978; Hem, 1985). The effect of this desorption and complexation process is evident by the relation between the two constituents, with the highest uranium concentrations occurring in water samples with bicarbonate concentrations more than 200 mg/L (fig. 10D).

Concentrations of radon-222 ranged from 95 to 3,600 pCi/L in the 20 samples, with a median of 261 pCi/L (fig. 10C and table 3). In samples having a sodium-bicarbonate water type, radon-222 concentrations ranged from 143 to 3,600 pCi/L, with a median of 395 pCi/L. Eight of the 12 samples having this water type, in addition to having pH values ranging from 8.0 to 8.7, exceeded the proposed MCL for radon-222 of 300 pCi/L (U.S. Environmental Protection Agency, 2012b). Because of the short half-life of 3.82 days for radon-222, high concentrations of this isotope in a groundwater sample indicate that a source of the immediate parent

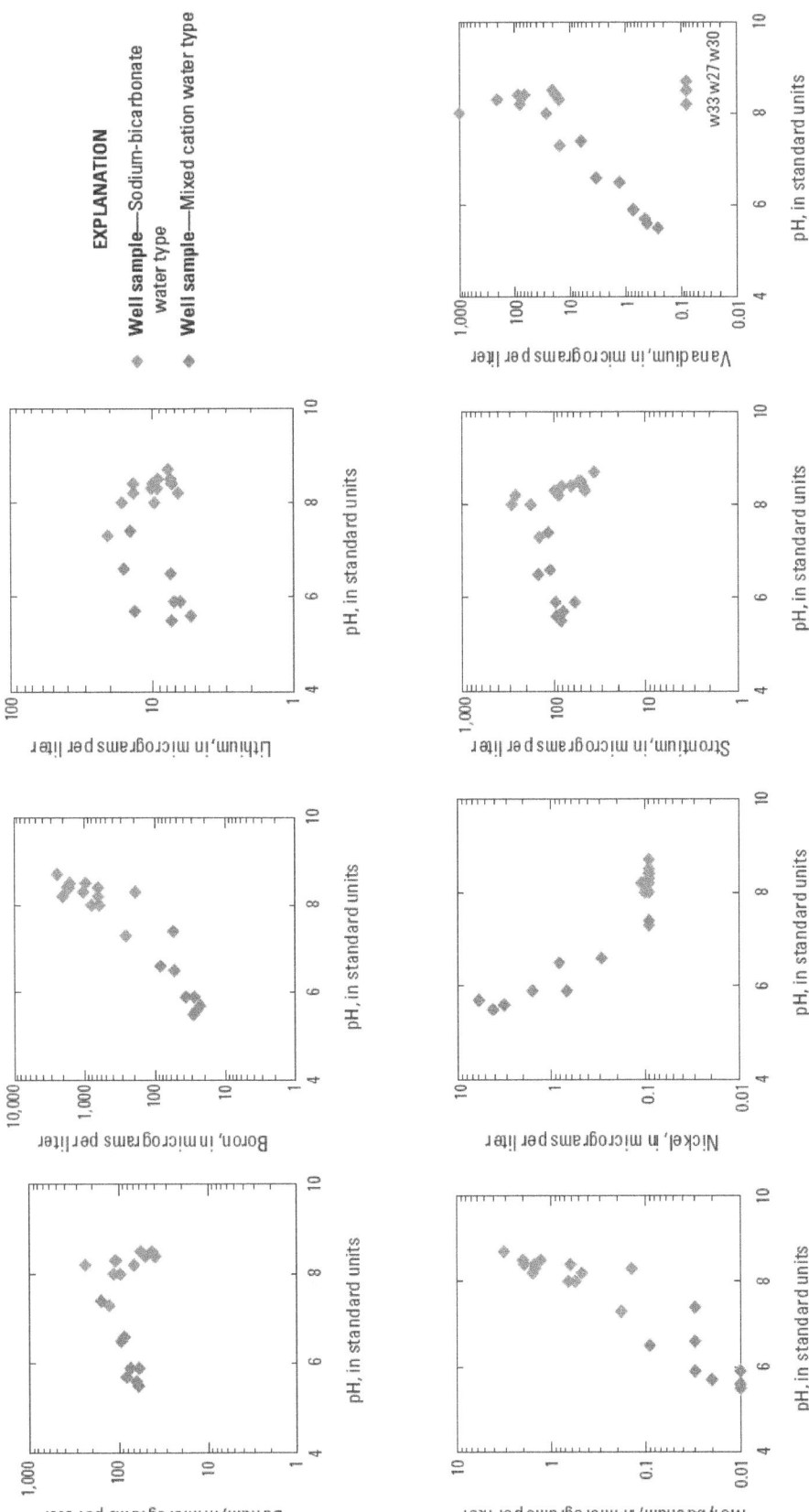

Figure 9. Concentrations of selected trace elements in water samples collected from 20 private wells in part of the Kickapoo Tribe of Oklahoma Jurisdictional Area, central Oklahoma, 2011.

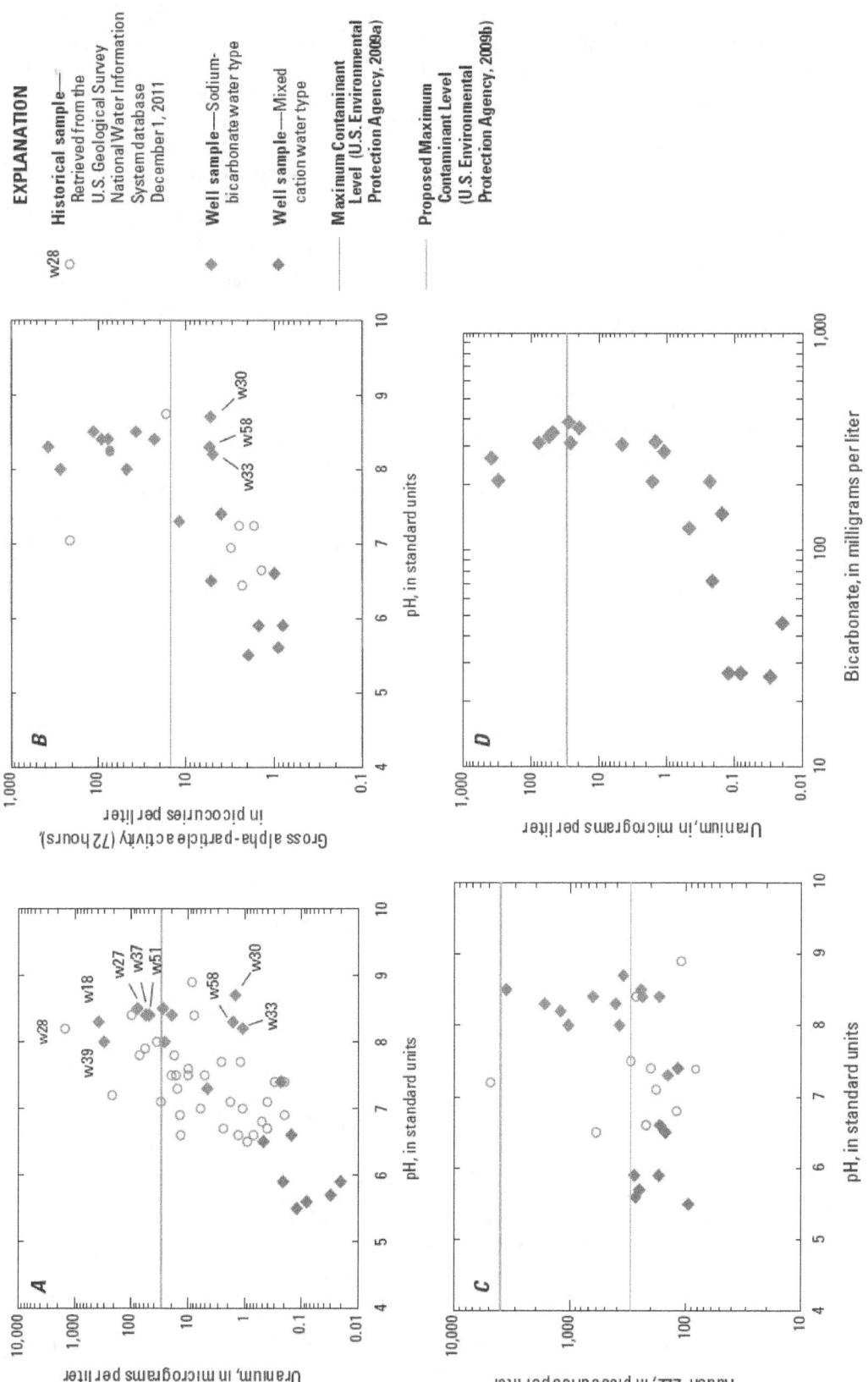

Figure 10. Concentrations of *A*, uranium, *B*, gross alpha-particle activity, and *C*, radon-222 in relation to pH and *D*, uranium relation to bicarbonate in water samples collected from 20 private wells for this study (2011) and in historical analyses (1977–2008) in the Kickapoo Tribe of Oklahoma Jurisdictional Area, central Oklahoma.

radionuclide radium-226, a product of the uranium-238 decay series, is present in aquifer materials near the well. Radon-222 concentrations in the eight samples having a mixed cation water type were lower ranging from 95 to 278 pCi/L, with a median of 168 pCi/L. Of 10 historical analyses for radon-222 in the tribal jurisdictional area, three groundwater samples had concentrations exceeding or equaling 300 pCi/L. One sample having a radon-222 concentration of 4,900 pCi/L also had an elevated uranium concentration of 217 µg/L and a nearly neutral pH of 7.2 (fig. 5C). Sodium was the dominant cation of this sample, but the lack of a bicarbonate concentration does not allow for the determination of water type.

In the study area, as in the rest of the Garber–Wellington aquifer, groundwater is predominately oxygenated (Parkhurst and others, 1996), and most pH values were above 6.0 (75 percent of 20 samples) (table 3 and app. 1), a geochemical condition which is not favorable for mobilizing radium from aquifer materials (Szabo and others, 2012). Concentrations of radium-226 and radium-228 (combined) in the 20 samples ranged from 0.03 to 1.7 pCi/L, with a median of 0.45 pCi/L, considerably less than the MCL of 5 pCi/L for combined radium. No historical samples for radium-226 in the tribal jurisdictional area were available, and of two samples analyzed for radium-228, each contained a concentration of 2 pCi/L or lower.

Both uranium-238 and radon-222 are alpha-particle emitters during the decay process; however, only uranium shows a clearly positive relation to gross alpha-particle activity (fig. 11) (radon is not shown, and it is important to note that, unlike uranium, radon is lost from samples by degassing before the gross alpha-particle activity measurement can be made). The relation between uranium concentration and gross alpha-particle activity indicates that a predominant source of the gross alpha-particle activity is uranium. Gross alpha-particle activities in samples with a sodium-bicarbonate water type ranged from 5 to 370 pCi/L (median 42 pCi/L) at 72 hours and from 4 to 350 pCi/L (median 41 pCi/L) at 30 days. Eight of the 12 samples that were a sodium-bicarbonate water type exceeded the 15 pCi/L MCL for gross alpha-particle activity in drinking water at 72 hours and at 30 days (w39, w18, w37, w47, w48, w51, w27, and w50) (app. 3 and fig. 11). These samples with a sodium-bicarbonate water type had pH values ranging from 8.0 to 8.5. Uranium concentrations in five of these samples exceeded the MCL of 30 µg/L, and the remaining three had uranium concentrations exceeding 10 µg/L (fig. 10A). Gross alpha-particle activity was substantially lower in samples of mixed cation water types with low uranium concentrations. Gross alpha-particle activities in mixed cation water type ranged from not detected (in one sample) to 5.2 pCi/L (median 1.2 pCi/L) at 72 hours and from not detected (in five samples) to 1 pCi/L at 30 days.

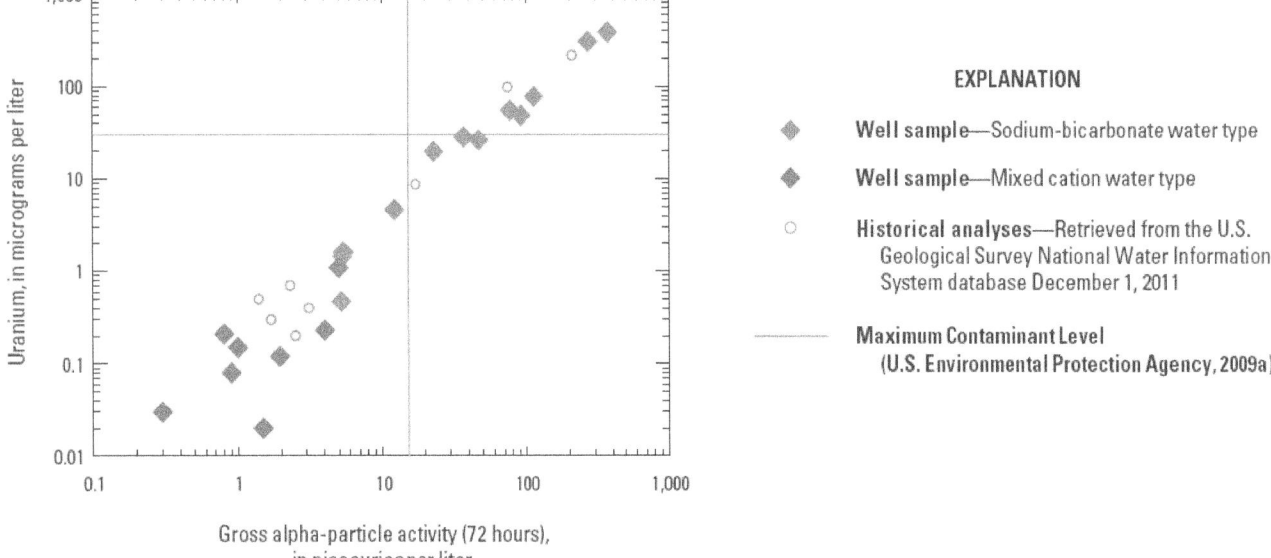

Figure 11.   Concentrations of uranium related to gross alpha-particle activity at 72 hours after collection in water samples from 20 private wells for this study (2011) and in historical analyses (1988–1990) in part of the Kickapoo Tribe of Oklahoma Jurisdictional Area, central Oklahoma.

Gross beta-particle activity also was higher in samples of sodium-bicarbonate water type than in samples of mixed cation water types and increased in both water types (15 of 20 samples) from 72 hours to 30 days. An increase in gross beta-particle activity over time probably results from ingrowth and decay of uranium daughter products that emit beta particles (Welch and others, 1995; Szabo and others, 2007). In sodium-bicarbonate water type, gross beta-particle activities ranged from not detected in three samples to 8.6 pCi/L at 72 hours (median 1.8 pCi/L) and from 1.8 to 102 pCi/L (median 8.1 pCi/L) at 30 days (app. 3 and table 3). Gross beta-particle activities in mixed cation water types were substantially less, ranging from not detected in two samples to 1.7 pCi/L at 72 hours and from 0.8 to 1.7 pCi/L at 30 days (median 1.2 pCi/L). The absence of an increase in beta-particle activity with time after sample collection is consistent with the low concentrations of uranium measured in the mixed cation water types.

Relation Between pH and Occurrence of Trace Elements and Radionuclides

Water-quality data collected during this study show that pH values above 8.0 are useful indicators for potentially high concentrations of uranium, radon-222, and gross alpha-particle activity in groundwater in the study area. This relation can be expected to be valid as long as the waters are also oxic. High pH values also are a useful indicator for potentially high concentrations of arsenic, chromium, and selenium in groundwater when these elements occur in the aquifer matrix along groundwater-flow paths. Values of pH were higher in wells completed in the Wellington Formation (of the Garber–Wellington aquifer), and deeper water had a higher potential for greater concentrations of uranium and the other oxyanions.

In general, concentrations of uranium and several other trace elements in groundwater in the study area are related to pH, oxic conditions, concentrations of the elements in the aquifer matrix, and groundwater-flow paths. The aquifer geochemistry tends to be strongly oxic, which supports these geochemical conditions even at substantial depths. The highest concentrations of uranium and other trace elements in the study area tended to occur in sodium-bicarbonate water type having pH values above 8.0. Concentrations of uranium and other trace elements in the aquifer rocks tended to increase with depth as the sediments become finer-grained in the Garber–Wellington aquifer (Gromadzki, 2004). It is likely that groundwater moves more slowly through these sediments, and with increased residence time, concentrations of dissolved solids (and specific conductance) increase. Shallow wells that produce young water of mixed cation water types having low pH values and low concentrations of dissolved solids appear to be less likely to contain high concentrations of radionuclides and trace elements and gross alpha- and beta-particle activities.

Summary

An assessment of groundwater quality was conducted by the U.S. Geological Survey during 2011, in cooperation with the Kickapoo Tribe of Oklahoma, to describe the occurrence of radionuclides and selected trace elements in groundwater and to determine if pH could be used as a surrogate for laboratory analyses to quickly and inexpensively identify wells that may contain relatively high concentrations of radionuclides. Residents of this rural area use groundwater from Quaternary-aged terrace deposits and the Permian-aged Garber–Wellington aquifer for domestic purposes. Measurements of pH and specific conductance from 59 private wells had a positive relation to land-surface altitude; water samples from shallow wells completed in the terrace deposits (at higher altitudes) tended to have lower pH values and lower dissolved solids concentrations than water samples from deeper wells completed in the Wellington Formation. Water samples from 20 wells, selected for chemical analyses, varied from a sodium-bicarbonate water type, characteristic of water with longer residence times in the aquifer having high pH values from cation exchange, to mixed cation water types, characteristic of more recently recharged water having lower pH values. Sodium and bicarbonate were the dominant cation and anion in samples from 12 of 20 wells sampled. Those 12 samples had high pH values ranging from 7.3 to 8.7 (median of 8.4) and concentrations of dissolved solids ranging from 248 to 501 milligrams per liter (mg/L) (median of 394 mg/L). In general, those 12 samples had the highest concentrations of selected trace elements and the radionuclides uranium and radon-222 and the highest gross alpha-particle activity.

Other trace elements that had a positive relation to pH were boron, molybdenum, and vanadium. Concentrations of barium, lithium, and strontium increased with pH but substantially decreased at pH values near and above 8.0, similar to calcium and magnesium. Elements that showed a negative relation to pH were nickel and beryllium, with concentrations of those trace elements being highest at pH values lower than 6.

Uranium concentrations ranged from 0.02 to 383 micrograms per liter (μg/L) (median of 1.54 μg/L) with the samples from 5 of the 20 wells exceeding the Maximum Contaminant Level (MCL) for public drinking-water supplies of 30 μg/L. The five wells with uranium concentrations exceeding 30 μg/L had pH values ranging from 8.0 to 8.5. Concentrations of radium-226 and radium-228 (combined) ranged from 0.03 to 1.7 picocuries per liter (pCi/L), well below the MCL of 5 pCi/L. Radon-222 concentrations ranged from 95 to 3,600 pCi/L (median of 261 pCi/L), with eight samples exceeding the proposed MCL of 300 pCi/L. Those eight samples had a sodium-bicarbonate water type and pH values ranging from 8.0 to 8.7. Eight samples exceeded the 15 pCi/L MCL for gross alpha-particle activity at 72 hours after collection and at 30 days after the initial count. Those samples were a sodium-bicarbonate water type and had pH values ranging from 8.0 to 8.5. All eight samples had uranium concentrations exceeding 10 μg/L and included the five samples exceeding the MCL of 30 μg/L indicating that uranium is the dominant source of the gross alpha-particle activity. Gross beta-particle activity in 15 of the 20 samples showed an increase from 72 hours to 30 days. An increase in gross beta-particle activity over time probably was caused by the ingrowth and decay of uranium daughter products that emit beta particles.

Water-quality data collected for this study indicate that pH values above 8.0 are a useful indicator for potentially high concentrations of uranium, gross alpha-particle activity, and radon-222, in the study area. The waters are oxic and contain abundant bicarbonate, both of which promote uranium solubility. Values of pH also are a useful indicator for potentially high concentrations of arsenic, chromium, and selenium in groundwater when these elements occur in the aquifer matrix along groundwater-flow paths

References

Al-Shaieb, Zuhair, Olmsted, R.W., Shelton, J.W., May, R.T., Owens, R.T., and Hanson, R.E., 1977, Uranium potential of Permian and Pennsylvanian sandstones in Oklahoma: American Association of Petroleum Geologists Bulletin, v. 61, no. 3, p. 360–375.

American Society of Testing and Materials International, 2010, Standard test method for radon in drinking water: West Conshohocken, Penn., ASTM International, p. 360–375.

Ayotte, J.D., Flanagan, S.M., and Morrow, W.S., 2007, Occurrence of uranium and 222radon in glacial and bedrock aquifers in the northern United States, 1993–2003: U.S. Geological Survey Scientific Investigations Report 2007–5037, 84 p. (Also available at http://pubs.er.usgs.gov/publication/sir20075037.)

Ayotte, J.D., Gronberg, J.M., and Apodaca, I.E., 2011, Trace elements and radon in groundwater across the United States, 1992–2003: U.S. Geological Survey Scientific Investigations Report 2011–5059, 115 p. (Also available at http://pubs.usgs.gov/sir/2011/5059.)

Back, William, 1966, Hydrochemical facies and ground-water flow patterns in the northern part of the Atlantic Coastal Plain: U.S. Geological Survey Professional Paper 498–A, 42 p. (Also available at http://pubs.er.usgs.gov/publication/pp498A.)

Becker, C.J., Smith, J.S., Greer, J.R. and Smith, Kevin, 2010, Arsenic-related water quality with depth and water quality of well-head samples from production wells, Oklahoma, 2008: U.S. Geological Survey Scientific Investigations Report 2010–5110, 46 p. (Also available at http://pubs.usgs.gov/sir/2010/5047/.)

Bingham, R.H., and Moore, R.L., 1975, Reconnaissance of the water resources of the Oklahoma City quadrangle, central Oklahoma: Oklahoma Geological Survey Hydrologic Atlas 4, 4 sheets. (Also available at http://www.ogs.ou.edu/pubsDLHAs.php.)

Breit, G.N., 1998, The diagenetic history of Permian rocks in the Central Oklahoma aquifer, in Christenson, Scott, and Havens, J.S., eds., Ground-water quality assessment of the Central Oklahoma aquifer, Oklahoma—Results of investigations: U.S. Geological Survey Water-Supply Paper 2357–A, p. 45–61. (Also available at http://pubs.er.usgs.gov/publication/wsp2357A.)

Childress, C.J.O., Forman, W.T., Connor, B.F., and Maloney, T.J., 1999, New reporting procedures based on long-term method detection levels and some considerations for interpretations of water-quality data provided by the U.S. Geological Survey National Water Quality Laboratory: U.S. Geological Survey Open-File Report 99–193, 19 p. (Also available at http://pubs.er.usgs.gov/publication/ofr99193#.)

Christenson, Scott, and Havens, J.S., eds., 1998, Ground-water-quality assessment of the Central Oklahoma aquifer, Oklahoma—Results of investigations: U.S. Geological Survey Water-Supply Paper 2357–A, 179 p. (Also available at http://pubs.er.usgs.gov/publication/wsp2357A.)

Christy, S.P., and Pope, J.P., 2009, High uranium concentrations in the Central Oklahoma aquifer near Guthrie, Logan County, Oklahoma [abs.]: Geological Society of America Abstracts with Programs, v. 41, no. 2, p. 9.

Commission on Life Sciences, 1999, Health effects of exposure to radon—BEIR VI (1999): Washington, D.C., National Academy Press. (Also available at http://www.nap.edu/openbook.php?record_id=5499&page=R1.)

Curtis, G.P., Fox, Patricia, Kohler, Matthias, and Davis, J.A., 2004, Comparison of in situ uranium KD values with a laboratory determined surface complexation model: Applied Geochemistry, v. 19, p. 1643–1653. (Also available at http://www.sciencedirect.com/science/article/pii/S0883292704000800.)

DeSimone, L.A., 2009, Quality of water from domestic wells in principal aquifers of the United States, 1991–2004: U.S. Geological Survey Scientific Investigations Report 2008–5227, 139 p. (Also available at http://pubs.usgs.gov/sir/2008/5227/.)

Driscoll, F.G., 1986, Groundwater and wells (2d ed.): St.Paul, Minn., Johnson Filtration Systems, Inc., 1,089 p.

Fishman, M.J., and Friedman, L.C., eds., 1989, Methods for determination of inorganic substances in water and fluvial sediments: U.S. Geological Survey Techniques of Water-Resources Investigations, book 5, chap. A1, 545 p. (Also available at http://pubs.er.usgs.gov/publication/twri05A1.)

Fishman, M.J., ed., 1993, Methods of analysis by the U.S. Geological Survey National Water Quality Laboratory—Determination of inorganic and organic constituents in water and fluvial sediments: U.S. Geological Survey Open-File Report 93–125, 217 p. (Also available at http://pubs.er.usgs.gov/publication/ofr93125.)

Focazio, M.J., Szabo, Zoltan, Kraemer, T.F., Mullin, A.H., Barringer, T.H., and dePaul, V.T., 2001, Occurrence of selected radionuclides in ground water used for drinking water in the United States—A reconnaissance survey, 1998: U.S. Geological Survey Water-Resources Investigations Report 00–4273, 39 p. (Also available at http://pubs.er.usgs.gov/publication/wri004273.)

Fry, Joyce, Xian, George, Jin, Suming, Dewitz, Jon, Homer, Collin, Yang, Limin, Barnes, Christopher, Herold, Nathaniel, and Wickham, James, 2011, Completion of the 2006 National Land Cover Database for the conterminous United States: Photogrammetric Engineering and Remote Sensing, v. 77, no. 9, p. 858–864. (Also available at http://www.mrlc.gov/nlcd2006.php.)

Garbarino, J.R., 1999, Methods of analysis by the U.S. Geological Survey National Water Quality Laboratory—Determination of dissolved arsenic, boron, lithium, selenium, strontium, thallium, and vanadium using inductively coupled plasma-mass spectrometry: U.S. Geological Survey Open-File Report 99–093, 31 p. (Also available at http://pubs.er.usgs.gov/publication/ofr9993.)

Garbarino, J.R., Kanagy, L.K., and Cree, M.E., 2006, Determination of elements in natural-water, biota, sediment and soil samples using collision/reaction cell inductively coupled plasma-mass spectrometry: U.S. Geological Survey Techniques and Methods, book 5, sec. B, chap. 1, 88 p. (Also available at http://pubs.er.usgs.gov/publication/tm5B1.)

Gromadzki, G.A., 2004, Outcrop-based gamma-ray characterization of arsenic-bearing lithofacies in the Garber–Wellington formation, Central Oklahoma aquifer, Cleveland County, Oklahoma: Stillwater, Oklahoma State University, M.S. thesis, 231 p.

Hall, F.R., Donahue, P.M., and Eldridge, A.L., 1985, Radon gas in ground water in New Hampshire: National Water Well Association Proceedings of the Second Annual Eastern Regional Ground Water Conference, Worthington, Ohio, p. 86–100.

Hem, J.D., 1985, Study and interpretation of the chemical characteristics of natural water (3d ed.): U.S. Geological Survey Water-Supply Paper 2254, 263 p. (Also available at http://pubs.er.usgs.gov/publication/wsp2254.)

Hodge, V.F., Stetzenbach, K.J., and Johannesson, K.H., 1998, Similarities in the chemical composition of carbonate ground waters and seawater: Environmental Science Technology, v. 32, p. 2481–2486.

Hsi, C.D., and Langmuir, Donald, 1985, Adsorption of uranyl onto ferric oxyhydroxides—Application of the surface-complexation site-binding model: Geochimica Cosmochimica Acta, v. 49, p. 1931–1941.

Jurgens, B.C., Fram, M.S., Belitz, Kenneth, Burow, K.R., and Landon, M.K., 2009, Effects of groundwater development on uranium—Central Valley, California, USA: Ground Water, v. 47, p. 1–16.

Langmuir, Donald, 1978, Uranium solution-mineral equilibria at low temperatures with applications to sedimentary ore deposits: Geochimica Cosmochimica Acta, v. 42, p. 547–569.

Lindberg, F.A., ed., 1987, Correlation of stratigraphic units of North America (COSUNA) project, Texas-Oklahoma tectonic region: American Association of Petroleum Geologists, 1 sheet.

Mashburn, S.L., and Magers, Jessica, 2011, Potentiometric surface in the Central Oklahoma (Garber–Wellington) aquifer, Oklahoma: U.S. Geological Survey, Scientific Investigations Map 3147, 1 sheet, scale 1:24,000. (Also available at http://pubs.usgs.gov/sim/3147/.)

McCurdy, D.E., Garbarino, J.R., and Mullin, A.H., 2008, Interpreting and reporting radiological water-quality data: U.S. Geological Survey Techniques and Methods, book 5, sec. B, chap. 6, 33 p. (Also available at http://pubs.usgs.gov/tm/05b06/.)

National Atmospheric Deposition Program, 2012, NADP/NTN Monitoring Location OK17. (Also available at http://nadp.sws.uiuc.edu/sites/siteinfo.asp?id=OK17&net=NTN.)

National Environmental Education Foundation, 2011, Acidic groundwater: accessed June 2011 at http://www.earthgauge.net/2009/acidic-groundwater.

National Environmental Methods Index, 1980, Prescribed procedures for measurement of radioactivity in drinking water—Gross alpha and gross beta radioactivity in drinking water: EPA 600/4-80-032, method ID 900.0. (Also available at https://www.nemi.gov/apex/f?p=237:38:135308361791914:::P38_METHOD_ID:4730.)

Oklahoma Water Resources Board, 2011, Multi-purpose well completion report database: accessed June 2011 at http://maps.owrb.state.ok.us/ms/ws/wellsbycounty.php.

Parkhurst, D.L., Christenson, Scott, and Breit, G.N., 1996, Ground-water quality assessment of the Central Oklahoma aquifer, Oklahoma—Geochemical and geohydrologic investigations: U.S. Geological Survey Water-Supply Paper 2357–C, 101 p. (Also available at http://pubs.er.usgs.gov/publication/wsp2357C.)

Piper, A.M., 1944, A graphic procedure in the geochemical interpretation of water analyses: Transactions, American Geophysical Union, v. 25, p. 914–923.

Plummer, L.N., and Back, William, 1980, The mass balance approach—Application to interpreting the chemical evolution of hydrologic systems: American Journal of Science, v. 280, no. 2, p. 130–142.

Robertson, F.N., 1989, Arsenic in ground-water under oxidizing conditions, south-west United States: Environmental Geochemistry and Health, v. 11, no. 3–4, p. 171–185. (Also available at http://www.springerlink.com/content/t36747742478t515/?MUD=MP.)

Rounds, S.A., and Wilde, F.D., eds., 2001, Alkalinity and acid neutralizing capacity (ver. 2.0): U.S. Geological Survey Techniques of Water-Resources Investigations, book 9, chap. A6, sec. 6.6. (Also available at http://pubs.water.usgs.gov/twri9A6/.)

Schlottmann, J.L., Mosier, E.L., and Breit, G.N., 1998, Arsenic, chromium, selenium, and uranium in the Central Oklahoma aquifer, in Christenson, Scott, and Havens, J.S., eds., Ground-water quality assessment of the Central Oklahoma aquifer, Oklahoma—Results of investigations: U.S. Geological Survey Water-Supply Paper 2357–A, p. 119–179. (Also available at http://pubs.er.usgs.gov/publication/wsp2357A.)

Smith, S.J., and Christenson, Scott, 2005, Naturally occurring arsenic in ground water, Norman, Oklahoma, 2004, and remediation options for produced water: U.S. Geological Survey Fact Sheet 2005–3111, 6 p. (Also available at http://pubs.er.usgs.gov/publication/fs20053111.)

Smith, S.J., Paxton, S.T., Christenson, Scott, Puls, R.W., and Greer, J.R., 2009, Flow contribution and water quality with depth in a test hole and public-supply wells—Implications for arsenic remediation through well modification, Norman, Oklahoma, 2003–2006: U.S. Environmental Protection Agency Office of Research and Development, National Risk Management Research Laboratory, 57 p. (Also available at http://cfpub.epa.gov/si/si_public_record_Report.cfm?dirEntryId=199062&CFID=89427770&CFTOKEN=54325115&jsessionid=cc30b92f2e0477bba92130163e63a123e127.)

Szabo, Zoltan, Fischer, J.M., dePaul, V.T., Jacobsen, Eric, and Kraemer, T.F., 2007, Change in gross alpha- and beta-particle activity after sample collection as an indicator of the occurrence of radium-224 and uranium in ground water in the Western United States, 2004: National Ground Water Association Arsenic, Radium, Radon, and Uranium Conference, Charleston, S.C., March 22–23, 2007, p. 3–4.

Szabo, Zoltan, Fischer, J.M., and Hancock, T.C., 2012, Principal aquifers can contribute radium to sources of drinking water under certain geochemical conditions: U.S. Geological Survey Fact Sheet 2010–3113, 6 p. (Also available at http://pubs.er.usgs.gov/publication/fs20103113.)

Tortorelli, R.L., 2009, Water use in Oklahoma 1950–2005: U.S. Geological Survey Scientific Investigations Report 2009–5212, 49 p. (Also available at http://pubs.er.usgs.gov/publication/sir20095212.)

U.S. Environmental Protection Agency, 1980, Prescribed procedures for measurement of radioactivity in drinking water, radium-228: EPA 600/4–80–032, method ID 904.0. (Also available at https://www.nemi.gov/apex/f?p=237:38:4400739314572075::::P38_METHOD_ID:4730.)

U.S. Environmental Protection Agency, 1999, Federal Guidance Report No. 13—Cancer risk coefficients for environmental exposure to radionuclides: Washington, D.C., EPA 402–R–99–001.

U.S. Environmental Protection Agency, 2000, National primary drinking water regulation; radionuclides; final rule: Federal Register, Dec. 7, 2000. (Also available at https://www.federalregister.gov/articles/2000/12/07/00-30421/national-primary-drinking-water-regulations-radionuclides-final-rule#h-28.)

U.S. Environmental Protection Agency, 2008a, Prescribed procedures for measurement of radioactivity in drinking water (Radon emanation technique), Radium-226, August 1980: EPA 600/4–80–032, method ID 903.1, (Also available at https://www.nemi.gov/apex/f?p=237:38:135308361791914:::P38_METHOD_ID:4732.)

U.S. Environmental Protection Agency, 2009a, Drinking-water contaminants—List of contaminants and their MCLs: accessed March 25, 2010, at http://www.epa.gov/safewater/contaminants/index.html.

U.S. Environmental Protection Agency, 2009b, National primary drinking water regulations—Radon-222: Federal Register Environmental Documents, v. 64, no. 211, p. 59, 245–59, 294. (Also available at http://www.epa.gov/fedrgstr/EPA-WATER/1999/November/Day-02/w27741a.htm.)

U.S. Environmental Protection Agency, 2010a, Radium: accessed March 25, 2010, at http://www.epa.gov/rpdweb00/radionuclides/radium.html.

U.S. Environmental Protection Agency, 2010b, Radon: accessed March 25, 2010, at http://www.epa.gov/rpdweb00/radionuclides/radon html.

U.S. Environmental Protection Agency, 2011, Uranium: accessed March 25, 2010, at http://www.epa.gov/rpdweb00/radionuclides/uranium html.

U.S. Environmental Protection Agency, 2012a, Beta particles: accessed March 25, 2010, at http://www.epa.gov/rpdweb00/understand/beta.html.

U.S. Environmental Protection Agency, 2012b, National primary drinking water regulations—Radon-222: accessed March 25, 2010, at http://www.epa.gov/fedrgstr/EPA-WATER/1999/November/Day-02/w27741a htm.

U.S. Environmental Protection Agency, 2012c, Secondary drinking water regulations—Guidance for nuisance chemicals: accessed March 25, 2010, at http://water.epa.gov/drink/contaminants/secondarystandards.cfm.

U.S. Geological Survey, 1987, Guidelines for sampling and analysis for dissolved radon-222 in ground water and surface water: U.S. Geological Survey, Office of Water Quality Technical Memorandum No. 88.02, October 13, 1987.

U.S. Geological Survey, 2006, Collection of water samples (ver. 2.0): U.S. Geological Survey Techniques of Water-Resources Investigations, book 9, chap. A4. (Also available at http://pubs.water.usgs.gov/twri9A4/.)

U.S. Geological Survey, 2012, National Water Information System Web interface—USGS water-quality data for Oklahoma: accessed December 1, 2011, at http://waterdata.usgs.gov/ok/nwis/qw.

Welch, A.H., Szabo, Zoltan, Parkhurst, D.L., Van Metre, P.C., and Mullin, A.H., 1995, Gross-beta activity in ground water—Natural sources and artifacts of sampling and laboratory analysis: Applied Geochemistry, v. 10, p. 491–503. (Also available at http://www.sciencedirect.com/science/article/pii/0883292795000208.)

Welch, A.H., Westjohn, D.B., Helsel, D.R., and Wanty, R.B., 2000, Arsenic in ground water of the United States—Occurrence and geochemistry: Ground Water, v. 38, no. 4, p. 589–604.

Wilde, F.D., ed., 2004, Cleaning of equipment for water sampling (ver. 2.0): U.S. Geological Survey Techniques of Water-Resources Investigations, book 9, chap. A3. (Also available at http://pubs.water.usgs.gov/twri9A3/.)

Wilde, F.D., and Radtke, D.B., eds., 1998, Methods for field measurements: U.S. Geological Survey Techniques of Water Resources Investigations, book 9, chap. A6.0–A6.6, 135 p. (Also available at http://pubs.water.usgs.gov/twri9A.)

Wilde, F.D., Radtke, D.B., Gibs, Jacob, and Iwatsubo, R.T., eds., 2004 (with updates through 2009), Processing of water samples (ver. 2.2): U.S. Geological Survey Techniques of Water-Resources Investigations, book 9, chap. A5. (Also available at http://pubs.water.usgs.gov/twri9A5/.)

Zapecza, O.S., and Szabo, Zoltan, 1988, Natural radioactivity in ground water—A review, in Moody, D.W., Carr, Jerry, Chase, E.B., and Paulson, R.W., comps., National water summary 1986—Hydrologic events and ground-water quality: U.S. Geological Survey Water-Supply Paper 2325, p. 50–57. (Also available at http://pubs.er.usgs.gov/publication/wsp2325.)

Appendixes 1–5

Appendix 1. Water-property measurements and concentrations of major ions measured in water samples collected from 20 private wells in part of the Kickapoo Tribe of Oklahoma Jurisdictional Area, central Oklahoma, 2011.

[ID, identifier; USGS, U.S. Geological Survey; mg/L, milligrams per liter; µS/cm, microsiemens per centimeter at 25 degrees Celsius; —, not measured or calculated; CaCO₃, calcium carbonate; SiO₂, silicon dioxide; E, estimated, mean titrant volume error indicates that substances in addition to bicarbonate were neutralized during titration; <, less than]

Well ID	USGS station ID	Sample date YYYYMMDD	Dissolved oxygen, in mg/L	pH, in standard units	Specific conductance, in µS/cm	Dissolved solids, sum, in mg/L	Calcium, in mg/L	Magnesium, in mg/L	Potassium, in mg/L	Sodium, in mg/L
w2	352844097034401	20110926	2.9	5.9	132	90	9.47	3.54	0.77	10.6
w3	352812097052101	20110822	6.7	7.4	346	194	33.8	14.2	1.22	17.7
w6	352813097032401	20110823	5.9	5.7	277	157	15.9	6.24	0.82	26.4
w17	352747097051401	20110913	5.9	6.6	285	161	19.7	9.11	1.22	26.9
w18	352717097043801	20111004	3.5	8.3	522	315	11.1	5.3	1.67	102
w20	352732097041501	20111012	6.4	5.9	184	110	15.3	6.43	0.68	11.6
w24	352726097031601	20111012	8.9	6.5	373	198	28.1	11.8	0.65	30.5
w27	352714097034101	20110822	0.8	8.5	690	401	2.58	1.39	0.96	155
w28[1]	352710097034601	20081203	1.6	8.2	443	[2]251	17.4	8.68	1.93	74.5
w30	352655097043201	20110830	2.2	8.7	870	501	1.94	0.838	0.88	193
w33	352642097034801	20110912	3.2	8.2	632	367	5.34	2.35	1.31	140
w37	352656097031601	20110928	0.8	8.4	687	409	4.42	2.13	1.1	156
w39	352654097022301	20110830	3.9	8	443	251	19.4	10.9	2.03	59.3
w41	352632097024401	20110831	5.9	5.5	156	89	11.1	4.42	0.93	9.37
w42	352625097022401	20110926	5.9	5.6	146	81	11.6	4.23	1.01	7.67
w47	352609097033801	20110830	3.2	8	559	330	8.95	5.61	1.97	113
w48	352603097024901	20110823	—	8.4	673	400	2.84	1.68	1	156
w50	352557097024101	20111005	1	8.5	809	485	3.13	1.83	1.15	184
w51	352553097021101	20110906	3.3	8.4	783	464	4.93	2.35	1.19	174
w56	352510097011201	20110831	4.3	7.3	693	389	28.5	14.5	1.5	103
w58	352417097015301	20110823	5.5	8.3	415	248	5.07	2.9	1.18	86.2

Well ID	USGS station ID	Date of sample	Alkalinity, in mg/L as CaCO$_3$	Bicarbonate, in mg/L	Bromide, in mg/L	Carbon dioxide, in mg/L	Carbonate, in mg/L	Chloride, in mg/L	Fluoride, in mg/L	Silica, in mg/L as SiO$_2$	Sulfate, in mg/L
w2	352844097034401	20110926	38	46	0.09	100	0	7.32	0.14	25.4	9.53
w3	352812097052101	20110822	170	207	0.05	13	0	6.89	0.16	14.6	2.71
w6	352813097032401	20110823	21	E26	0.31	93	0	49.1	<0.04	20.4	25.1
w17	352747097051401	20110913	121	147	0.1	58	0	8.73	0.16	16.4	6.11
w18	352717097043801	20111004	222	265	0.1	2.1	3	26	0.7	11.2	21.8
w20	352732097041501	20111012	59	72	0.11	150	0	9.63	0.27	19.1	11.1
w24	352726097031601	20111012	104	126	0.31	69	0	23.8	0.28	16	23.8
w27	352714097034101	20110822	263	312	0.12	1.6	5	30.9	1	8.86	40.5
w28[1]	352710097034601	20081203	227	275	0.06	2.9	1	4.67	0.24	10.3	10.8
w30	352655097043201	20110830	273	315	0.23	1.1	8	76.5	1.23	8.56	51.5
w33	352642097034801	20110912	238	284	0.12	2.8	3	29.7	0.82	9.69	33.1
w37	352656097031601	20110928	281	334	0.1	2.2	4	27.1	0.85	10	36.7
w39	352654097022301	20110830	173	209	0.11	3.6	0	16.1	0.36	12.9	24.1
w41	352632097024401	20110831	22	27	0.13	133	0	15.8	0.04	15.5	18.5
w42	352625097022401	20110926	22	27	0.09	118	0	11.7	0.04	13.4	17.6
w47	352609097033801	20110830	258	311	0.05	5.5	2	12.6	0.49	12.6	19.1
w48	352603097024901	20110823	308	365	0.07	2.1	5	16.1	0.57	11.9	24.3
w50	352557097024101	20111005	328	388	0.15	2.1	6	41.4	0.72	11.2	44.4
w51	352553097021101	20110906	291	347	0.11	2	4	28.8	0.89	9.11	66.2
w56	352510097011201	20110831	252	306	0.23	23	0	40.6	0.47	15.3	33.7
w58	352417097015301	20110823	173	207	0.07	1.7	2	17.9	0.25	16.6	13.3

[1]Well was sampled and water properties measured by the USGS in December 2008. Well was subsequently destroyed.

[2]Dissolved solids of filtered water dried at 180 degrees Celsius.

Appendix 2. Trace-element concentrations measured in water samples collected from 20 private wells in part of the Kickapoo Tribe of Oklahoma Jurisdictional Area, central Oklahoma, 2011.

[ID, identifier; USGS, U.S. Geological Survey; µg/L, micrograms per liter; <, less than ; E, estimated, result is below the laboratory reporting level and above the long-term method detection level]

Well ID	USGS station ID	Sample date	Antimony, in µg/L	Arsenic, in µg/L	Barium, in µg/L	Beryllium, in µg/L	Boron, in µg/L	Bromide, in mg/L	Cadmium, in µg/L	Chromium, in µg/L	Cobalt, in µg/L	Copper, in µg/L	Iron, in µg/L
w2	352844097034401	20110926	<0.03	0.15	60	0.04	28	0.09	<0.02	0.52	<0.02	0.74	<3
w3	352812097052101	20110822	<0.03	0.62	159	<0.01	56	0.05	<0.02	2.7	<0.02	<0.50	<3
w6	352813097032401	20110823	<0.03	0.16	83	0.25	23	0.31	<0.02	0.61	0.08	0.87	6
w17	352747097051401	20110913	<0.03	0.63	88	0.01	85	0.1	<0.02	0.9	0.02	<0.50	<3
w18	352717097043801	20111004	0.05	6.6	108	0.01	1,050	0.1	<0.02	3.2	<0.02	<0.80	<3
w20	352732097041501	20111012	<0.03	0.22	75	0.04	37	0.11	<0.02	1.2	<0.02	<0.80	3
w24	352726097031601	20111012	<0.03	0.23	95	0.02	54	0.31	<0.02	1	0.03	<0.80	<3
w27	352714097034101	20110822	<0.03	6.8	57	0.02	1,630	0.12	<0.02	<0.06	<0.02	<0.50	<3
w28[1]	352710097034601	20081203	E.04	3.6	239	<0.02	647	0.06	<0.02	0.24	E0.01	2.3	<4
w30	352655097043201	20110830	<0.03	0.15	45	0.04	2,460	0.23	<0.02	<0.06	<.02	<0.50	<3
w33	352642097034801	20110912	<0.03	0.54	68	0.04	2,050	0.12	<0.02	<0.06	0.02	<0.50	<3
w37	352656097031601	20110928	<0.03	4.8	51	0.02	1,670	0.1	<0.02	4.5	<0.02	<0.50	<3
w39	352654097022301	20110830	0.05	24.7	97	0.01	802	0.11	<0.02	1.3	0.02	1.6	3
w41	352632097024401	20110831	<0.03	0.07	61	0.18	29	0.13	<0.02	0.26	0.05	21.6	3
w42	352625097022401	20110926	<0.03	0.1	64	0.08	26	0.09	<0.02	0.12	0.06	0.62	7
w47	352609097033801	20110830	<0.03	0.33	115	0.01	624	0.05	<0.02	24.4	<0.02	0.8	4
w48	352603097024901	20110823	<0.03	0.57	40	0.01	651	0.07	<0.02	31.4	<0.02	<0.50	<3
w50	352557097024101	20111005	<0.03	0.72	43	0.01	979	0.15	<0.02	15.6	<0.02	<0.80	<3
w51	352553097021101	20110906	<0.03	3.1	39	0.02	1,760	0.11	<0.02	4.7	<0.02	<0.50	<3
w56	352510097011201	20110831	<0.03	0.73	130	<0.01	262	0.23	<0.02	11.2	<0.02	<0.50	<3
w58	352417097015301	20110823	<0.03	0.9	110	<0.01	192	0.07	<0.02	8.6	<0.02	0.51	<3

Well ID	USGS station ID	Sample date	Lead, in µg/L	Lithium, in µg/L	Manganese, in µg/L	Molybdenum, in µg/L	Nickel, in µg/L	Selenium, in µg/L	Silver, in µg/L	Strontium, in µg/L	Thallium, in µg/L	Vanadium, in µg/L	Zinc, in µg/L
w2	352844097034401	20110926	0.39	7.1	<0.1	<0.01	1.6	1.1	<0.01	58.5	<0.01	0.73	48.5
w3	352812097052101	20110822	0.03	14.4	<0.1	0.03	<0.09	0.23	<0.01	115	<0.01	6.3	<1.4
w6	352813097032401	20110823	0.06	13.4	5.3	0.02	5.9	0.15	<0.01	78.9	<0.01	0.45	71.8
w17	352747097051401	20110913	0.03	16	0.7	0.03	0.29	0.36	<0.01	110	<0.01	3.4	<1.4
w18	352717097043801	20111004	<0.03	9.3	<0.1	1.49	<0.09	0.76	<0.01	98.1	<0.01	207	3.4
w20	352732097041501	20111012	0.03	6.4	0.4	0.03	0.68	0.3	<0.01	95.4	<0.01	0.74	<1.4
w24	352726097031601	20111012	0.06	7.5	<0.1	0.09	0.82	1.2	<0.01	149	<0.01	1.3	<1.4
w27	352714097034101	20110822	0.05	7.5	1.5	2	<0.09	<0.03	0.01	53.1	<0.01	<0.08	<1.4
w28[1]	352710097034601	20081203	<0.06	13.6	0.7	0.47	E0.11	5.1	<0.01	262	0.04	81	<2.0
w30	352655097043201	20110830	0.16	7.8	3.1	3.16	<0.09	<0.03	<0.01	36.2	<0.01	<0.08	<1.4
w33	352642097034801	20110912	<0.01	6.6	4.3	1.57	<0.09	<0.03	0.01	87.8	<0.01	<0.08	<1.4
w37	352656097031601	20110928	<0.01	7.3	<0.1	1.51	<0.09	147	<0.01	64.2	<0.01	86.4	<1.4
w39	352654097022301	20110830	0.07	9.7	<0.1	0.65	0.1	36.8	<0.01	288	<0.01	1,000	3.6
w41	352632097024401	20110831	0.14	7.4	0.4	<0.01	4.2	0.2	<0.01	82.8	<0.01	0.26	2.3
w42	352625097022401	20110926	0.06	5.4	0.6	<0.01	3.2	0.22	<0.01	94.3	<0.01	0.41	1.7
w47	352609097033801	20110830	<0.01	16.5	<0.1	0.55	<0.09	24	<0.01	178	<0.01	27	1.5
w48	352603097024901	20110823	0.18	13.7	<0.1	0.62	<0.09	12.8	<0.01	48	<0.01	18	<1.4
w50	352557097024101	20111005	<0.03	9.2	<0.1	1.29	<0.09	29.3	<0.01	50.1	<0.01	20.9	<1.4
w51	352553097021101	20110906	<0.01	10	0.3	1.93	<0.09	23.1	<0.01	80.7	<0.01	66.9	<1.4
w56	352510097011201	20110831	0.03	20.8	<0.1	0.18	<0.09	4.4	<0.01	144	<0.01	15.2	<1.4
w58	352417097015301	20110823	0.03	10.2	<0.1	0.14	<0.09	2.8	<0.01	45.2	<0.01	15.6	1.6

[1]Well was sampled and water properties measured by the USGS in December 2008. Well was subsequently destroyed.

Appendix 3. Radionuclide concentrations measured in water samples collected from 20 private wells in part of the Kickapoo Tribe of Oklahoma Jurisdictional Area, central Oklahoma, 2011.

[ID, identifier; USGS, U.S. Geological Survey; Result, radiological concentrations plus or minus the 1-sigma combined standard uncertainty; GA (30d), sample used for the 72-hour gross alpha-particle analysis is counted a second time approximately 30 days after the initial count as referenced to a detector calibrated by using 230Thorium; ND, analyte not detected (concentration is less than the sample-specific critical level); pCi/L, picocurie per liter; GA (72h), sample analyzed for gross alpha-particle activity at approximately 72 hours after sample collection as referenced to a detector calibrated by using 230Thorium; GB (30d), sample used for the 72-hour gross beta-particle analysis is counted a second time approximately 30 days after the initial count as referenced to a detector calibrated by using 137Cesium; GB (72h), sample analyzed for gross beta-particle activity at approximately 72 hours after sample collection as referenced to a detector calibrated by using 137Cesium; µg/L, micrograms per liter; D, diluted sample, method high range exceeded; B, laboratory background blank greater than the sample-specific critical level; C, sample-specific critical level exceeded the minimum detectable concentration for a sample; L, laboratory control sample recovery is outside acceptable range; ±, plus or minus]

Well ID	USGS station ID	Sample date YYYYMMDD	Sample time	Radiological constituent	Result	Sample-specific critical level	Remark	Units	Sample type
w2	352844097034401	20110926	1430	GA (30d)	-0.3 ± 0.26	0.48	ND	pCi/L	Filtered.
w2	352844097034401	20110926	1430	GA (72h)	1.5 ± 0.51	0.51		pCi/L	Filtered.
w2	352844097034401	20110926	1430	GB (30d)	0.9 ± 0.4	0.6		pCi/L	Filtered.
w2	352844097034401	20110926	1430	GB (72h)	0.8 ± 0.4	0.61		pCi/L	Filtered.
w2	352844097034401	20110926	1430	Radium-226	0.076 ± 0.015	0.018		pCi/L	Filtered.
w2	352844097034401	20110926	1430	Radium-228	0.25 ± 0.081	0.19		pCi/L	Filtered.
w2	352844097034401	20110926	1430	Radon-222	170 ± 15	13.8		pCi/L	Unfiltered.
w2	352844097034401	20110926	1430	Uranium	0.02	0.004		µg/L	Filtered.
w3	352812097052101	20110822	1345	GA (30d)	0.9 ± 0.66	0.85		pCi/L	Filtered.
w3	352812097052101	20110822	1345	GA (72h)	4 ± 0.73	0.51		pCi/L	Filtered.
w3	352812097052101	20110822	1345	GB (30d)	1.7 ± 0.43	0.61		pCi/L	Filtered.
w3	352812097052101	20110822	1345	GB (72h)	1.6 ± 0.31	0.42		pCi/L	Filtered.
w3	352812097052101	20110822	1345	Radium-226	0.21 ± 0.027	0.015		pCi/L	Filtered.
w3	352812097052101	20110822	1345	Radium-228	0.49 ± 0.094	0.2		pCi/L	Filtered.
w3	352812097052101	20110907	1030	Radon-222	117 ± 11	10.7		pCi/L	Unfiltered.
w3	352812097052101	20110822	1345	Uranium	0.23	0.004		µg/L	Filtered.
w6	352813097032401	20110823	0915	GA (30d)	0.3 ± 0.42	0.59	ND	pCi/L	Filtered.
w6	352813097032401	20110823	0915	GA (72h)	0.3 ± 0.31	0.45	ND	pCi/L	Filtered.
w6	352813097032401	20110823	0915	GB (30d)	1.4 ± 0.43	0.62		pCi/L	Filtered.
w6	352813097032401	20110823	0915	GB (72h)	0.3 ± 0.28	0.44	ND	pCi/L	Filtered.
w6	352813097032401	20110823	0915	Radium-226	0.042 ± 0.012	0.015		pCi/L	Filtered.
w6	352813097032401	20110823	0915	Radium-228	-0.01 ± 0.095	0.23	ND	pCi/L	Filtered.
w6	352813097032401	20110906	1130	Radon-222	251 ± 18	12.1		pCi/L	Unfiltered.
w6	352813097032401	20110823	0915	Uranium	0.03	0.004		µg/L	Filtered.
w17	352747097051401	20110913	1100	GA (30d)	0.8 ± 0.5	0.59		pCi/L	Filtered.
w17	352747097051401	20110913	1100	GA (72h)	1 ± 0.42	0.53	B	pCi/L	Filtered.

Well ID	USGS station ID	Sample date YYYYMMDD	Sample time	Radiological constituent	Result	Sample-specific critical level	Remark	Units	Sample type
w17	352747097051401	20110913	1100	GB (30d)	1.7 ± 0.44	0.62		pCi/L	Filtered.
w17	352747097051401	20110913	1100	GB (72h)	1.2 ± 0.29	0.42		pCi/L	Filtered.
w17	352747097051401	20110913	1100	Radium-226	0.15 ± 0.021	0.018		pCi/L	Filtered.
w17	352747097051401	20110913	1100	Radium-228	0.4 ± 0.1	0.23		pCi/L	Filtered.
w17	352747097051401	20110913	1100	Radon-222	167 ± 14	10.9		pCi/L	Unfiltered.
w17	352747097051401	20110913	1100	Uranium	0.15	0.004		µg/L	Filtered.
w18	352717097043801	20111004	1400	GA (30d)	350 ± 36	0.85		pCi/L	Filtered.
w18	352717097043801	20111004	1400	GA (72h)	370 ± 38	0.47		pCi/L	Filtered.
w18	352717097043801	20111004	1400	GB (30d)	102 ± 5.8	0.67		pCi/L	Filtered.
w18	352717097043801	20111004	1400	GB (72h)	6.9 ± 0.71	0.48		pCi/L	Filtered.
w18	352717097043801	20111004	1400	Radium-226	0.77 ± 0.069	0.021		pCi/L	Filtered.
w18	352717097043801	20111004	1400	Radium-228	0.68 ± 0.13	0.26		pCi/L	Filtered.
w18	352717097043801	20111004	1400	Radon-222	1,680 ± 93	12.2		pCi/L	Unfiltered.
w18	352717097043801	20111004	1400	Uranium	383	0.004	D	µg/L	Filtered.
w20	352732097041501	20111012	1000	GA (30d)	0.2 ± 0.28	0.42	ND	pCi/L	Filtered.
w20	352732097041501	20111012	1000	GA (72h)	0.8 ± 0.43	0.51		pCi/L	Filtered.
w20	352732097041501	20111012	1000	GB (30d)	0.8 ± 0.3	0.45		pCi/L	Filtered.
w20	352732097041501	20111012	1000	GB (72h)	0.3 ± 0.39	0.61	ND	pCi/L	Filtered.
w20	352732097041501	20111012	1000	Radium-226	0.07 ± 0.013	0.018		pCi/L	Filtered.
w20	352732097041501	20111012	1000	Radium-228	0.17 ± 0.09	0.2	ND	pCi/L	Filtered.
w20	352732097041501	20111012	1000	Radon-222	278 ± 19	11.6		pCi/L	Unfiltered.
w20	352732097041501	20111012	1000	Uranium	0.21	0.004		µg/L	Filtered.
w24	352726097031601	20111012	1330	GA (30d)	1 ± 0.49	0.63		pCi/L	Filtered.
w24	352726097031601	20111012	1330	GA (72h)	5.2 ± 1.1	0.9		pCi/L	Filtered.
w24	352726097031601	20111012	1330	GB (30d)	1.7 ± 0.41	0.6		pCi/L	Filtered.
w24	352726097031601	20111012	1330	GB (72h)	1.7 ± 0.56	0.85		pCi/L	Filtered.
w24	352726097031601	20111012	1330	Radium-226	0.095 ± 0.014	0.016		pCi/L	Filtered.
w24	352726097031601	20111012	1330	Radium-228	0.42 ± 0.088	0.19		pCi/L	Filtered.
w24	352726097031601	20111012	1330	Radon-222	149 ± 13	11.6		pCi/L	Unfiltered.
w24	352726097031601	20111012	1330	Uranium	0.47	0.004		µg/L	Filtered.
w27	352714097034101	20110822	1030	GA (30d)	116 ± 13	1.4	C	pCi/L	Filtered.
w27	352714097034101	20110822	1030	GA (72h)	114 ± 12	1.1		pCi/L	Filtered.
w27	352714097034101	20110822	1030	GB (30d)	22 ± 1.7	1		pCi/L	Filtered.
w27	352714097034101	20110822	1030	GB (72h)	1.6 ± 0.56	0.75		pCi/L	Filtered.

Appendix 3. Radionuclide concentrations measured in water samples collected from 20 private wells in part of the Kickapoo Tribe of Oklahoma Jurisdictional Area, central Oklahoma, 2011.—Continued

[ID, identifier; USGS, U.S. Geological Survey; Result, radiological concentrations plus or minus the 1-sigma combined standard uncertainty; GA (30d), sample used for the 72-hour gross alpha-particle analysis is counted a second time approximately 30 days after the initial count as referenced to a detector calibrated by using 230Thorium; ND, analyte not detected (concentration is less than the sample-specific critical level); pCi/L, picocurie per liter; GA (72h), sample analyzed for gross alpha-particle activity at approximately 72 hours after sample collection as referenced to a detector calibrated by using 230Thorium; GB (30d), sample used for the 72-hour gross beta-particle analysis is counted a second time approximately 30 days after the initial count as referenced to a detector calibrated by using 137Cesium; GB (72h), sample analyzed for gross beta-particle activity at approximately 72 hours after sample collection as referenced to a detector calibrated by using 137Cesium; µg/L, micrograms per liter; D, diluted sample; method high range exceeded; B, laboratory background blank greater than the sample-specific critical level; C, sample-specific critical level exceeded the minimum detectable concentration for a sample; L, laboratory control sample recovery is outside acceptable range; ±, plus or minus]

Well ID	USGS station ID	Sample date YYYYMMDD	Sample time	Radiological constituent	Result	Sample-specific critical level	Remark	Units	Sample type
w27	352714097034101	20110822	1030	Radium-226	0.43 ± 0.042	0.016		pCi/L	Filtered.
w27	352714097034101	20110822	1030	Radium-228	0.29 ± 0.15	0.24		pCi/L	Filtered.
w27	352714097034101	20110906	1100	Radon-222	3,600 ± 200	12		pCi/L	Unfiltered.
w27	352714097034101	20110822	1030	Uranium	77.9	0.004		µg/L	Filtered.
w28	352710097034601	20081203	1030	Uranium	1,500	0.006	D	µg/L	Filtered.
w30	352655097043201	20110830	1430	GA (30d)	4 ± 1.1	1		pCi/L	Filtered.
w30	352655097043201	20110830	1430	GA (72h)	5.3 ± 1.3	1.2	B	pCi/L	Filtered.
w30	352655097043201	20110830	1430	GB (30d)	1.5 ± 0.61	0.91		pCi/L	Filtered.
w30	352655097043201	20110830	1430	GB (72h)	0.8 ± 0.65	1	ND	pCi/L	Filtered.
w30	352655097043201	20110830	1430	Radium-226	0.18 ± 0.023	0.017		pCi/L	Filtered.
w30	352655097043201	20110830	1430	Radium-228	0.33 ± 0.12	0.24	L	pCi/L	Filtered.
w30	352655097043201	20110906	1000	Radon-222	350 ± 23	12		pCi/L	Unfiltered.
w30	352655097043201	20110830	1430	Uranium	1.46	0.004		µg/L	Filtered.
w33	352642097034801	20110912	1230	GA (30d)	11.1 ± 1.7	0.82		pCi/L	Filtered.
w33	352642097034801	20110912	1230	GA (72h)	5 ± 1.1	0.93	B	pCi/L	Filtered.
w33	352642097034801	20110912	1230	GB (30d)	3.5 ± 0.54	0.71		pCi/L	Filtered.
w33	352642097034801	20110912	1230	GB (72h)	2.8 ± 0.52	0.73		pCi/L	Filtered.
w33	352642097034801	20110912	1230	Radium-226	1.2 ± 0.1	0.016		pCi/L	Filtered.
w33	352642097034801	20110912	1230	Radium-228	0.53 ± 0.11	0.23		pCi/L	Filtered.
w33	352642097034801	20110912	1230	Radon-222	1,220 ± 69	11.9		pCi/L	Unfiltered.
w33	352642097034801	20110912	1230	Uranium	1.09	0.004		µg/L	Filtered.
w37	352656097031601	20110928	1430	GA (30d)	80 ± 8.9	1		pCi/L	Filtered.
w37	352656097031601	20110928	1430	GA (72h)	77 ± 8.6	1.1		pCi/L	Filtered.
w37	352656097031601	20110928	1430	GB (30d)	16.4 ± 1.3	1.2		pCi/L	Filtered.
w37	352656097031601	20110928	1430	GB (72h)	1.9 ± 0.66	0.92		pCi/L	Filtered.
w37	352656097031601	20110928	1430	Radium-226	-0.01 ± 0.01	0.021	ND	pCi/L	Filtered.

Well ID	USGS station ID	Sample date YYYYMMDD	Sample time	Radiological constituent	Result	Sample-specific critical level	Remark	Units	Sample type
w37	352656097031601	20110928	1430	Radium-228	0.48 ± 0.089	0.2		pCi/L	Filtered.
w37	352656097031601	20110928	1430	Radon-222	640 ± 38	11		pCi/L	Unfiltered.
w37	352656097031601	20110928	1430	Uranium	55.1	0.004		µg/L	Filtered.
w39	352654097022301	20110830	0900	GA (30d)	260 ± 27	0.66		pCi/L	Filtered.
w39	352654097022301	20110830	0900	GA (72h)	270 ± 29	0.71	B	pCi/L	Filtered.
w39	352654097022301	20110830	0900	GB (30d)	89 ± 5	0.57		pCi/L	Filtered.
w39	352654097022301	20110830	0900	GB (72h)	8.6 ± 0.8	0.57		pCi/L	Filtered.
w39	352654097022301	20110830	0900	Radium-226	0.69 ± 0.065	0.019		pCi/L	Filtered.
w39	352654097022301	20110830	0900	Radium-228	0.47 ± 0.11	0.25	L	pCi/L	Filtered.
w39	352654097022301	20110907	1130	Radon-222	1,040 ± 59	10.9		pCi/L	Unfiltered.
w39	352654097022301	20110830	0900	Uranium	306	0.004	D	µg/L	Filtered.
w41	352632097024401	20110831	1200	GA (30d)	0.4 ± 0.36	0.47	ND	pCi/L	Filtered.
w41	352632097024401	20110831	1200	GA (72h)	2 ± 0.42	0.36		pCi/L	Filtered.
w41	352632097024401	20110831	1200	GB (30d)	1 ± 0.4	0.6		pCi/L	Filtered.
w41	352632097024401	20110831	1200	GB (72h)	1.3 ± 0.29	0.43		pCi/L	Filtered.
w41	352632097024401	20110831	1200	Radium-226	0.133 ± 0.019	0.018		pCi/L	Filtered.
w41	352632097024401	20110831	1200	Radium-228	0.14 ± 0.11	0.2	ND	pCi/L	Filtered.
w41	352632097024401	20110831	1200	Radon-222	95 ± 11	13.1		pCi/L	Unfiltered.
w41	352632097024401	20110831	1200	Uranium	0.12	0.004		µg/L	Filtered.
w42	352625097022401	20110926	1145	GA (30d)	-0.2 ± 0.29	0.49	ND	pCi/L	Filtered.
w42	352625097022401	20110926	1145	GA (72h)	0.9 ± 0.36	0.33		pCi/L	Filtered.
w42	352625097022401	20110926	1145	GB (30d)	0.9 ± 0.38	0.58		pCi/L	Filtered.
w42	352625097022401	20110926	1145	GB (72h)	1.2 ± 0.4	0.59		pCi/L	Filtered.
w42	352625097022401	20110926	1145	Radium-226	0.04 ± 0.01	0.013		pCi/L	Filtered.
w42	352625097022401	20110926	1145	Radium-228	0.22 ± 0.086	0.2		pCi/L	Filtered.
w42	352625097022401	20110926	1145	Radon-222	270 ± 20	13.9		pCi/L	Unfiltered.
w42	352625097022401	20110926	1145	Uranium	0.08	0.004		µg/L	Filtered.
w47	352609097033801	20110830	1230	GA (30d)	43 ± 5	0.97		pCi/L	Filtered.
w47	352609097033801	20110830	1230	GA (72h)	47 ± 5.4	0.81	B	pCi/L	Filtered.
w47	352609097033801	20110830	1230	GB (30d)	8.8 ± 0.73	0.65		pCi/L	Filtered.
w47	352609097033801	20110830	1230	GB (72h)	2.4 ± 0.49	0.64		pCi/L	Filtered.
w47	352609097033801	20110830	1230	Radium-226	0.23 ± 0.028	0.018		pCi/L	Filtered.
w47	352609097033801	20110830	1230	Radium-228	0.2 ± 0.1	0.23	L	pCi/L	Filtered.
w47	352609097033801	20110906	0900	Radon-222	380 ± 25	11.9		pCi/L	Unfiltered.

Appendix 3. Radionuclide concentrations measured in water samples collected from 20 private wells in part of the Kickapoo Tribe of Oklahoma Jurisdictional Area, central Oklahoma, 2011.—Continued

[ID, identifier; USGS, U.S. Geological Survey; Result, radiological concentrations plus or minus the 1-sigma combined standard uncertainty; GA (30d), sample used for the 72-hour gross alpha-particle analysis is counted a second time approximately 30 days after the initial count as referenced to a detector calibrated by using 230Thorium; ND, analyte not detected (concentration is less than the sample-specific critical level); pCi/L, picocurie per liter; GA (72h), sample analyzed for gross alpha-particle activity at approximately 72 hours after sample collection as referenced to a detector calibrated by using 230Thorium; GB (30d), sample used for the 72-hour gross beta-particle analysis is counted a second time approximately 30 days after the initial count as referenced to a detector calibrated by using 137Cesium; GB (72h), sample analyzed for gross beta-particle activity at approximately 72 hours after sample collection as referenced to a detector calibrated by using 137Cesium; μg/L, micrograms per liter; D, diluted sample; method high range exceeded; B, laboratory background blank greater than the sample-specific critical level; C, sample-specific critical level exceeded the minimum detectable concentration for a sample; L, laboratory control sample recovery is outside acceptable range; ±, plus or minus]

Well ID	USGS station ID	Sample date YYYYMMDD	Sample time	Radiological constituent	Result	Sample-specific critical level	Remark	Units	Sample type
w47	352609097033801	20110830	1230	Uranium	26.3	0.004		μg/L	Filtered.
w48	352603097024901	20110823	1130	GA (30d)	24 ± 3.7	1.6		pCi/L	Filtered.
w48	352603097024901	20110823	1130	GA (72h)	23 ± 3	0.77		pCi/L	Filtered.
w48	352603097024901	20110823	1130	GB (30d)	6 ± 1	1.4		pCi/L	Filtered.
w48	352603097024901	20110823	1130	GB (72h)	1 ± 0.65	1	ND	pCi/L	Filtered.
w48	352603097024901	20110823	1130	Radium-226	0.031 ± 0.0089	0.013		pCi/L	Filtered.
w48	352603097024901	20110823	1130	Radium-228	0.04 ± 0.11	0.26	ND	pCi/L	Filtered.
w48	352603097024901	20110906	0900	Radon-222	169 ± 14	11.8		pCi/L	Unfiltered.
w48	352603097024901	20110823	1130	Uranium	19.7	0.004		μg/L	Filtered.
w50	352557097024101	20111005	1030	GA (30d)	39 ± 4.7	1.4		pCi/L	Filtered.
w50	352557097024101	20111005	1030	GA (72h)	37 ± 4.5	1.3		pCi/L	Filtered.
w50	352557097024101	20111005	1030	GB (30d)	7.4 ± 0.85	1		pCi/L	Filtered.
w50	352557097024101	20111005	1030	GB (72h)	1.3 ± 0.65	1		pCi/L	Filtered.
w50	352557097024101	20111005	1030	Radium-226	0.03 ± 0.01	0.011		pCi/L	Filtered.
w50	352557097024101	20111005	1030	Radium-228	0.3 ± 0.12	0.26		pCi/L	Filtered.
w50	352557097024101	20111005	1030	Radon-222	244 ± 19	13.8		pCi/L	Unfiltered.
w50	352557097024101	20111005	1030	Uranium	28.2	0.004		μg/L	Filtered.
w51	352553097021101	20110906	1430	GA (30d)	74 ± 8.3	1.3		pCi/L	Filtered.
w51	352553097021101	20110906	1430	GA (72h)	92 ± 10	0.85		pCi/L	Filtered.
w51	352553097021101	20110906	1430	GB (30d)	12 ± 1	0.94		pCi/L	Filtered.
w51	352553097021101	20110906	1430	GB (72h)	0.3 ± 0.6	0.87	ND	pCi/L	Filtered.
w51	352553097021101	20110906	1430	Radium-226	0.086 ± 0.015	0.015		pCi/L	Filtered.
w51	352553097021101	20110906	1430	Radium-228	0.15 ± 0.11	0.22	ND	pCi/L	Filtered.
w51	352553097021101	20110906	1430	Radon-222	237 ± 18	12		pCi/L	Unfiltered.
w51	352553097021101	20110906	1430	Uranium	48.3	0.004		μg/L	Filtered.
w56	352510097011201	20110831	1400	GA (30d)	9 ± 2	1.6	C	pCi/L	Filtered.

Well ID	USGS station ID	Sample date YYYYMMDD	Sample time	Radiological constituent	Result	Sample-specific critical level	Remark	Units	Sample type
w56	352510097011201	20110831	1400	GA (72h)	12 ± 1.8	0.98		pCi/L	Filtered.
w56	352510097011201	20110831	1400	GB (30d)	2.4 ± 0.81	1.2		pCi/L	Filtered.
w56	352510097011201	20110831	1400	GB (72h)	1.4 ± 0.56	0.84		pCi/L	Filtered.
w56	352510097011201	20110831	1400	Radium-226	0.094 ± 0.017	0.014		pCi/L	Filtered.
w56	352510097011201	20110831	1400	Radium-228	0.2 ± 0.1	0.2	ND	pCi/L	Filtered.
w56	352510097011201	20110913	1230	Radon-222	143 ± 12	11		pCi/L	Unfiltered.
w56	352510097011201	20110831	1400	Uranium	4.61	0.004		µg/L	Filtered.
w58	352417097015301	20110823	1400	GA (30d)	5.7 ± 1.2	0.91		pCi/L	Filtered.
w58	352417097015301	20110823	1400	GA (72h)	5.4 ± 0.94	0.64		pCi/L	Filtered.
w58	352417097015301	20110823	1400	GB (30d)	1.8 ± 0.49	0.71		pCi/L	Filtered.
w58	352417097015301	20110823	1400	GB (72h)	2.3 ± 0.34	0.46		pCi/L	Filtered.
w58	352417097015301	20110823	1400	Radium-226	0.4 ± 0.043	0.017		pCi/L	Filtered.
w58	352417097015301	20110823	1400	Radium-228	0.35 ± 0.13	0.25		pCi/L	Filtered.
w58	352417097015301	20111012	1100	Radon-222	410 ± 26	11.6		pCi/L	Unfiltered.
w58	352417097015301	20110823	1400	Uranium	1.62	0.004		µg/L	Filtered.

Appendix 4.   Analytical relative percent difference for concentrations of major ions and trace elements (including uranium) in quality-control samples, in addition to equipment-blank results, for water samples collected from 20 private wells in part of the Kickapoo Tribe of Oklahoma Jurisdictional Area, central Oklahoma, 2011.

[ID, identifier; mg/L, milligrams per liter; SiO2, silicon dioxide; RPD, relative percent difference; <, less than; µg/L, micrograms per liter; —, not calculated; RPD was not calculated if one constituent had an estimated concentration or a concentration less than the minimum reporting level]

Well ID	Sample date YYYYMMDD	Sample type	Calcium, in mg/L	Magnesium, in mg/L	Potassium, in mg/L	Sodium, in mg/L	Chloride, in mg/L	Fluoride, in mg/L	Silica, as SiO2, in mg/L	Sulfate, in mg/L
w41	20110831	Environmental	11.1	4.42	0.93	9.37	15.8	0.04	15.5	18.5
w41	20110831	Replicate	11.3	4.5	0.9	9.27	15.8	0.06	15.4	18.6
RPD			1.79 percent	1.79 percent	3.3 percent	1.1 percent	0	40 percent	0.7 percent	0.5 percent
w33	20110912	Environmental	5.34	2.35	1.31	140	29.7	0.82	9.69	33.1
w33	20110912	Replicate	5.37	2.35	1.27	141	29.7	0.81	9.62	32.7
RPD			0.6 percent	0	3.1 percent	0.7 percent	0	1.2 percent	0.7 percent	1.2 percent
		Equipment blank	<0.022	<0.011	<0.03	<0.06	<0.06	<0.04	<0.018	<0.09

Well ID	Sample date YYYYMMDD	Sample type	Antimony, in µg/L	Arsenic, in µg/L	Barium, in µg/L	Beryllium, in µg/L	Boron, in µg/L	Bromide, in µg/L	Cadmium, in µg/L	Chromium, in µg/L	Cobalt, in µg/L
w41	20110831	Environmental	<0.03	0.07	61	0.18	29	0.13	<0.02	0.26	0.05
w41	20110831	Replicate	<0.03	0.08	62	0.18	29	0.13	<0.02	0.25	0.05
RPD			—	13.3 percent	1.6 percent	—	—	0	—	3.9 percent	0
w33	20110912	Environmental	<0.03	0.54	68	0.04	2,050	0.12	<0.02	<0.06	0.02
w33	20110912	Replicate	<0.03	0.56	69	0.04	2,020	0.09	<0.02	<0.06	<0.02
RPD			—	3.64 percent	1.46 percent	0	1.47 percent	28.6 percent	—	—	0
		Equipment blank	<0.027	<0.03	<0.07	<0.006	<3	<0.01	<0.016	<0.07	<0.021

Well ID	Sample date YYYYMMDD	Sample type	Copper, in µg/L	Iron, in µg/L	Lead, in µg/L	Lithium, in µg/L	Manganese, in µg/L	Molybdenum, in µg/L	Nickel, in µg/L	Selenium, in µg/L	Silver, in µg/L
w41	20110831	Environmental	21.6	3	0.14	7.4	0.4	<0.01	4.2	0.2	<0.01
w41	20110831	Replicate	21.6	4	0.14	7.4	0.5	<0.01	4.2	0.17	<0.01
RPD			0	28.6 percent	0	0	22.2 percent	—	0	16.2 percent	—
w33	20110912	Environmental	<0.5	<3	<0.01	6.6	4.3	1.57	<0.09	<0.03	0.01
w33	20110912	Replicate	<0.5	5	<0.01	6.8	4.3	1.57	<0.09	<0.03	<0.01
RPD			—	—	—	3 percent	0	0	—	—	—
	Equipment blank		<0.8	<3.2	<0.025	<0.22	<0.13	<0.014	<0.09	<0.03	<0.005

Well ID	Sample date YYYYMMDD	Sample type	Strontium, in µg/L	Thallium, in µg/L	Uranium, in µg/L	Vanadium, in µg/L	Zinc, in µg/L
w41	20110831	Environmental	82.8	<0.01	0.08	0.26	2.3
w41	20110831	Replicate	83.2	<0.01	0.12	0.27	2.3
RPD			0.5 percent	—	40 percent	3.8 percent	0
w33	20110912	Environmental	87.8	<0.01	1.09	<0.08	<1.4
w33	20110912	Replicate	87.8	<0.01	1.11	<0.08	<1.4
RPD			0	—	1.8 percent	—	—
	Equipment blank		<0.2	<0.01	<0.004	<0.08	<1.4

Appendix 5. Analytical relative percent difference for concentrations of radionuclides (not including uranium) in quality-control samples for water samples collected from 20 private wells in part of the Kickapoo Tribe of Oklahoma Jurisdictional Area, central Oklahoma, 2011.

[ID, identifier; Result, radiological concentrations plus or minus the 1-sigma combined standard uncertainty; RPD, relative percent difference; RPD was not calculated if one constituent was not detected; —, not calculated; pCi/L, picocurie per liter; GA (72h), sample analyzed for gross alpha-particle activities at approximately 72 hours after sample collection as referenced to a detector calibrated by using 230Thorium; B, laboratory background blank greater than the sample-specific critical level; GA (30d), sample used for the 72-hour gross alpha-particle analysis is counted a second time approximately 30 days after the initial count as referenced to a detector calibrated by using 230Thorium; ND, not detected; GB (72h), sample analyzed for gross beta-particle activities at approximately 72 hours after sample collected as referenced to a detector calibrated by using 137Cesium; GB (30d), sample used for the 72-hour gross beta-particle analysis is counted a second time approximately 30 days after the initial count as referenced to a detector calibrated by using 137Cesium; ±, plus or minus]

Well ID	Quality control sample type	Sample date YYYYMMDD	Radiological constituent	Result	Sample-specific critical level	Remark	RPD	Units
w41	Environmental	20110831	Radium-226	0.133±0.019	0.018		13.6 percent	pCi/L
w41	Replicate	20110831	Radium-226	0.116±0.018	0.016			pCi/L
w33	Environmental	20110912	Radium-226	1.2±0.1	0.016		11.4 percent	pCi/L
w33	Replicate	20110912	Radium-226	1.07±0.097	0.016			pCi/L
w41	Environmental	20110831	GA (72h)	2±0.42	0.36		16.2 percent	pCi/L
w41	Replicate	20110831	GA (72h)	1.7±0.35	0.24			pCi/L
w33	Environmental	20110912	GA (72h)	5±1.1	0.93	B	41.3 percent	pCi/L
w33	Replicate	20110912	GA (72h)	7.6±1.4	0.95	B		pCi/L
w41	Environmental	20110831	GA (30d)	0.4±0.36	0.47		—	pCi/L
w41	Replicate	20110831	GA (30d)	0±0.31	0.49	ND		pCi/L
w33	Environmental	20110912	GA (30d)	11.1±1.7	0.82		0.9 percent	pCi/L
w33	Replicate	20110912	GA (30d)	11±1.7	0.82			pCi/L
w41	Environmental	20110831	GB (72h)	1.3±0.29	0.43		16.7 percent	pCi/L
w41	Replicate	20110831	GB (72h)	1.1±0.33	0.5			pCi/L
w33	Environmental	20110912	GB (72h)	2.8±0.52	0.73		11.3 percent	pCi/L
w33	Replicate	20110912	GB (72h)	2.5±0.57	0.81			pCi/L
w41	Environmental	20110831	GB (30d)	1±0.4	0.6		—	pCi/L
w41	Replicate	20110831	GB (30d)	0.6±0.44	0.7	ND		pCi/L

Well ID	Quality control sample type	Sample date YYYYMMDD	Radiological constituent	Result	Sample-specific critical level	Remark	RPD	Units
w33	Environmental	20110912	GB (30d)	3.5±0.54	0.71		2.9 percent	pCi/L
w33	Replicate	20110912	GB (30d)	3.4±0.52	0.68			pCi/L
w41	Environmental	20110831	Radium-228	0.14±0.11	0.2		60 percent	pCi/L
w41	Replicate	20110831	Radium-228	0.26±0.096	0.19			pCi/L
w33	Environmental	20110912	Radium-228	0.53±0.11	0.23		18.6 percent	pCi/L
w33	Replicate	20110912	Radium-228	0.44±0.11	0.23			pCi/L
w41	Environmental	20110831	Radon-222	95±11	13.117		8.1 percent	pCi/L
w41	Replicate	20110831	Radon-222	103±12	13.2773			pCi/L
w33	Environmental	20110912	Radon-222	1,220±69	11.9307		3.2 percent	pCi/L
w33	Replicate	20110912	Radon-222	1,260±71	12.0823			pCi/L

www.ingramcontent.com/pod-product-compliance
Lightning Source LLC
Chambersburg PA
CBHW081618170526
45166CB00009B/3022